NORTH CAROLINA

DEPARTMENT OF CONSERVATION AND DEVELOPMENT

R. BRUCE ETHERIDGE, *Director*

BULLETIN NUMBER 38

GOLD DEPOSITS

IN

NORTH CAROLINA

BY

HERMAN J. BRYSON

State Geologist

RALEIGH

1936

Table of Contents

Fig. 1. Flotation Mill, Rudisil Mine, Charlotte, N. C.

LETTER OF TRANSMITTAL

Raleigh, North Carolina,
March 1, 1936.

To his Excellency, HON. J. C. B. EHRINGHAUS,

Governor of North Carolina

Sir: I have the honor to submit herewith, as Bulletin No. 38, a report on Gold Deposits in North Carolina. Since there have been so many requests for the information contained in this bulletin, the Department has long felt the need of such a report.

The report summarizes the old publications relating to gold deposits in North Carolina, has brought this information up to date, and includes reports, information, and recent developments in the gold mining industry.

Yours respectfully,

R. BRUCE ETHERIDGE,
Director.

The following report is not an original work but a compilation of old reports now out of print and no longer available for distribution. The writer has endeavored to describe the various gold-bearing localities as fully as is thought practicable.

The information includes a brief description of each mine, the nature and type of ore deposits, and the processes used in extracting the gold.

The writer wishes to acknowledge thanks for the valuable assistance rendered by W. L. Cotton, Albemarle, N. C., for accompanying him to the various mines in Stanly, Montgomery, Cabarrus, Rowan, and Union counties; to acknowledge thanks to A. L. Nash, Salisbury, N. C., for information on the Harkey and Snyder mines, in Cabarrus County; to W. A. White, Charlotte, N. C., for information on the Rudisil Mine and flotation mill, in Mecklenburg County; to J. P. Sloss, Gold Hill, N. C., for information on mines in Cabarrus and Rowan counties; to E. L. Hertzog, Spartanburg, S. C., for information on the Snyder Mine; to E. M. Scott, New London, N. C., for information on the Parker Mine; to the Cassidy Brothers, Albemarle, N. C., for information on the Crowell Mine; to J. C. Lettellar, Wood, N. C., and W. L. Long, Raleigh, N. C., for information on the Portis Mine. All of these men spared neither time nor effort in placing information in the hands of the writer; and without this assistance it would have been impossible to have secured full information on the gold mines.

HERMAN J. BRYSON.

GOLD DEPOSITS IN NORTH CAROLINA

CHAPTER I

HISTORICAL NOTES ON GOLD MINING IN NORTH CAROLINA

The earliest reports relating to gold mining in North Carolina show evidence that gold mining was carried on within the borders of the State previous to the Revolutionary War. However, no absolutely authentic reference can be obtained; so the date of actual discovery of gold in North Carolina will always remain an uncertainty. Available information states that gold was produced in North Carolina at the Oliver Mine in Gaston County, the Dunn Mine in Mecklenburg County, and the Parker Mine in Cherokee County previous to the Revolutionary War. However, for North Carolina the first mint returns appear in 1793, but the first authentic record of gold being discovered in North Carolina is the finding of a 17-pound nugget on the Reed plantation in Cabarrus County, in 1799.

Below is a statement taken from John H. Wheeler's "Historical Sketches of North Carolina, 1584 to 1851."

"The first piece of gold found at this mine, was in the year 1799, by Conrad Reed, a boy of about twelve years old, a son of John Reed, the proprietor. The discovery was made in an accidental manner. The boy above named, in company with a sister and younger brother, went to a small stream, called Meadow Creek, on a Sabbath day, while their parents were at church, for the purpose of shooting fish with bow and arrow, and while engaged along the bank of the creek, Conrad saw a yellow substance shining in the water. He went in and picked it up, and found it to be some kind of metal, and carried it home. Mr. Reed examined it, but as gold was unknown in this part of the country at that time, he did not know what kind of metal it was: the piece was about the size of a small smoothing iron.

"Mr. Reed carried the piece of metal to Concord, and showed it to a

William Atkinson, a silversmith, but he not thinking of gold, was unable to say what kind of metal it was.

"Mr. Reed kept the piece for several years on his house floor, to lay against the door to keep it from shutting. In the year 1802, he went to market to Fayetteville, and carried the piece of metal with him, and on showing it to a jeweller, the jeweller immediately told him it was gold, and requested Mr. Reed to leave the metal with him and said he would flux it. Mr. Reed left it, and returned in a short time, and on his return the jeweller showed him a large bar of gold, six or eight inches long. The jeweller then asked Mr. Reed what he would take for the bar. Mr. Reed, not knowing the value of gold, thought he would ask a "big price" and so he asked three dollars and fifty cents ($3.50)! The jeweller paid him his price.

"After returning home, Mr. Reed examined and found gold in the surface along the creek. He then associated Frederick Kisor, James Love, and Martin Phifer with himself, and in the year 1803, they found a piece of gold in the branch that weighed twenty-eight pounds. Numerous pieces were found at this mine weighing from sixteen pounds down to the smallest particles. The whole surface along the creek for nearly a mile was very rich in gold.

"The veins of this mine were discovered in the year 1831. They yielded a large quantity of gold. The veins are flint or quartz.

"I do certify that the foregoing is a true statement of the discovery and history of this mine, as given by John Reed and his son Conrad Reed, now both dead.

January, 1848   GEORGE BARNHARDT.

"Weight of different pieces of gold found at this mine:

1803,	28 lbs.	1824,	16	lbs.
1804,	9 "	"	9½	"
"	7 "	"	8	"
"	3 "	1835	13½	"
"	2 "	"	4½	"
"	1¾ lbs.	"	4	"
		"	1	"
		"	8	"

115 lbs. steelyard weight.

"The annual products of the gold mines of the State, have been estimated at five hundred thousand dollars.* The produce of Cabarrus mines in 1840, by the census was estimated at thirty-five hundred dollars."

Later reports add to the above statement that Mr. John Reed, after learning the real value of gold, went back to the jeweller and recovered about $3,000. Since the publication of the above statement 38 pounds in nuggets were found at the Reed mine, making a total of 153 pounds in nuggets.

After additional nuggets were discovered at the Reed plantation, a systematic search was made for gold throughout the southeastern section of the United States.

By 1825 gold mining was carried on rather vigorously, and it was about 1833 that the greatest activity was known in this State. The gold placers in Burke and McDowell counties were first worked in 1828, and these placers were later traced southwestward into Georgia. The most extensive gravel deposits in the South Mountains district were on the headwaters of the First and Second Broad River, Muddy Creek, and Silver Creek, in the counties of Rutherford, McDowell, and Burke. This area embraced about 200 square miles.

Very soon after the discovery of gold in the South Mountains region, slave owners found a new and profitable use for their slaves.

In 1926, while in conference with a Mr. Kelly, who lived near Dysartville, and who was 84 years of age at the time, he stated to me that he had seen negro slaves go to work in the placer deposits in the morning and return at night with a quart jar full of gold. He stated that the slave owners would require from 4 to 8 slaves to fill the quart jar every day.

Old geological reports state that as many as 3,000 slaves

---

*Report by John H. Wheeler, Superintendent of Branch Mint, to the Secretary of the Treasury, in 1838.

could be seen working the gravel deposits along a single stream at the height of activity in this section. At first, most of the activity was along the streams, especially Silver Creek, but later spread to the hillsides. The hillsides contained gold in bench gravels as well as in the upper decomposed layer of country rock. This latter type of gold deposits was formed by secular disintegration and drift. The amount of gold produced in this manner is not known, and it is impossible even to approximate the amount, but the best authorities place the amount in excess of two or three million dollars.

At this particular time there was only one mint established by the government and this was at Philadelphia. Due to the long distance to the mint and inadequate means of transportation, it was only natural that the miners would look for other means of converting their gold into a circulating medium.

Three years after the discovery of gold, or about the time of the greatest production, 1831, a German, by the name of Christian Bechtler, a jeweller by trade, and living about three miles from Rutherfordton, proposed to the miners that he coin their gold for a small percentage. Since this saved considerable time and expense in transporting the gold to Philadelphia, the miners immediately accepted Bechtler's proposition. A large quantity of the gold was coined in $1, $2.50, and $5 pieces with the name of "C. Bechtler, Rutherford County, North Carolina," on one side, and on the reverse side, the value, number of grains, and carats fine. It is reported that the Federal Government instituted an investigation but in all instances found that the assay value of the coins was equal and in some instances greater than the denominational value; therefore the circulation was not disturbed. The percentage charged by Bechtler for coinage was 2½ per cent. Bechtler continued this coinage until his death in 1843, after which his nephew, C. Bechtler,

Jr., continued until June, 1857. From 1831 to 1843, it is reported that $4,000 to $5,000 were coined in a week, and the annual quantity was about equal for a period of 10 years. It is therefore impossible to estimate with any degree of accuracy the amount of gold produced in the South Mountains section of North Carolina previous to the Civil War.

The Bechtler coins today command a premium. The $2.50 pieces have sold recently at prices ranging from $10 to $20 each. These coins today are scarce and are seldom found except when owned by the older inhabitants of that section.

In the earlier gold mining activities in North Carolina the principle source of gold was from the placers and the decomposed rock near the surface, as well as to some extent from the upper decomposed part of the vein. Since so much profit had been made from the placer operations, considerable search was made for vein and lode deposits. By 1850, several important mines had been opened, especially the Reed Mine, in 1828; the Gold Hill Mine, in 1842; the Phifer, Davis, and Pewter mines in Union County; the Hearne Mine, Stanly County; the Long Creek and Reynolds mines in Montgomery County; the Kings Mountain Mine in Gaston County; the Phoenix and Barnhardt mines in Cabarrus County; the Rymer (Reimer) and Fisher Hills mines in Rowan County; the Rudisil Mine in Mecklenburg County; the Conrad Hill and Silver Hill mines in Davidson County. These mines operated with varying success until the Civil War.

With the discovery of gold in California in 1849, a great many of the prospectors, miners, and men experienced in mining migrated to the newly discovered region. This left a great many of the mines in North Carolina without experienced personnel; as a result a great many of the mines in North Carolina closed. Some of the more profitable mines, however, continued until the Civil War, at which time all mines closed for a period of a few years.

In the days of reconstruction a great many of the mines, such as the Gold Hill, Rymer, Rudisil, and Capps mines were reopened. New mines were also discovered in Union, Montgomery, and Mecklenburg counties. Among the more important in Montgomery County were the Coggins and Iola mines. The most important periods of production of gold in North Carolina were the periods 1831-1843, 1882-1891, 1902-1906, 1912-1915.

The present period of activity began in 1933. Several old mines have been unwatered and further examinations made. The most important operations undertaken at the present time, and which apparently are proving successful, are the Rudisil Mine, Mecklenburg County; the Howie Mine, Union County; the Whitney-Isenhour group in Rowan and Cabarrus counties; and the Portis Mine in Franklin County. In addition to these properties, considerable prospecting and development work are being carried on at the Snyder, Allen-Furr, and Harkey mines in Cabarrus County; the Parker and Thompson mines in Stanly County; the Young and Conrad Hill mines in Davidson County; and several placer properties in Montgomery, Randolph, and Rutherford counties.

CHAPTER II

DEVELOPMENT OF GOLD DEPOSITS

Mining, recovery, and production; character and fineness of gold.

GENERAL STATEMENT

Soon after the discovery of gold in Cabarrus County in 1799, that is when the real identity of the gold was known, in 1802, a vigorous search was made for other deposits in the southeastern part of Cabarrus County. In the beginning the principal search for gold was confined largely to the search for nuggets such as had been encountered along the streams on the Reed plantation.

The search for nuggets was continued for some time with varying success; but as time went on, nugget gold was hard to find. In this search for the nugget gold, however, the miners discovered that a great deal of fine gold was found in the sands and gravels in which the larger nuggets were discovered. With the discovery of this fine gold in the gravels, sands, and clay, the question arose as to how this gold should be recovered. Eventually, varying methods of recovery were evolved until more or less complicated methods were developed.

PLACER AND SOFT OXIDIZED ORES

It is not definitely known what primitive washing methods were first used in North Carolina, but, as has been in newly discovered regions since that time, the first washing was probably done with the pan. This primitive washing method was used for some time, but as the workings grew more extensive this method was superseded by the hand rocker, long tom, and sluice box. Even today these primitive methods still survive, and in a great many sections of North Carolina large numbers of men can be seen searching for placer material along the streams and hillsides, as well as searching for veins which have not heretofore been discovered.

The pan, sluice box, hand rockers, and long toms were soon superseded by the log washer, disintegrator, and Snodgrass machines. These latter machines were used especially in sections where the gold occurs in the tough tenacious clays and upper decomposed layers of the country rock. The revolving drum or blind trommel, into which the tough clay had been placed, with a certain amount of quartz pebbles, has taken the place, in a great many localities, of the log washers and other types of disintegrators. Within recent years these revolving steel drums, blind as well as open trommels, have been very successful in disintegrating the tougher clays. These trommels are especially satisfactory where an abundance of water is to be had.

Probably one of the most interesting methods of working the gravels and clays is that method first applied at Brindletown in the South Mountains district in 1883. This was known as the *hydraulic gravel elevator*. This method was especially satisfactory in sections where the fall was less than 1 foot in 100 feet. The level condition was a feature common to many of the southern placers; so, in order to overcome this obstacle for hydraulicking with continuous sluice, the hydraulic gravel elevator was decided upon.

In 1843-4 the first dredge was attempted in North Carolina in the Catawba River section of Gaston County. The sands and gravels in the river were scooped up by men using long-handled scoops, the material placed on flat boats and carried to the shore to be washed. Very soon after this type of dredging, mechanical dredges were introduced. These dredges were tried at the Portis Mine in addition to the sections along the Catawba River. A Bucyrus-Erie placer dredge No. 97, with buckets of 3 cubic feet capacity, owned by the Uharie River Gold Mining and Dredging Company, was used for some time on Uharie River in Montgomery County. It did not prove successful, however.

FREE MILLING, HARD ORES

The early miners, in their search for veins, were rewarded by the discovery of a number of veins in Cabarrus and Stanly counties. With the discovery of the veins, especially as the ores became harder, other methods of recovery had to be resorted to.

The first account of vein mining is in 1825 at the Barringer Mine, Stanly County. The upper soft decomposed part of the vein was washed in hand rockers and other simple devices, and considerable success was the result. However, as the miners went down on the veins the hand rockers and other primitive methods were not satisfactory. Some sort of crushing was necessary. Probably the first crushing method resorted to was with the hand mortar and subsequent panning. This type of crushing was carried on, however, for some time by the native tributors. Since this method was slow and the miners were not rewarded sufficiently for their efforts, other types of crushing were developed. The simplest type of mill was the *drag mill* or *arrastra*. The drag mill was soon followed by the introduction of the *Chilean mill* and eventually the *stamp mill*. The drag (arrastra) and Chilean mills were probably drawn from South American and Mexican practice and were probably the first mechanical pulverizing machines used in America in gold mining operations. Sometime after the introduction of the stamp mill, a type of mill more or less on the same principle of the Chilean mill was the *Lane mill*.

As time went on, a combination of the various methods of crushing were used. The soft ores were usually crushed in the drag and Chilean mills, while the harder ore was sent to the stamp mill.

During the past five-year period the old Chilean and

Lane mills, long out of use, have been reconditioned and put into operation. However, apparently none of these mills were a financial success, as they were soon abandoned. The stamp mill, however, is one of the chief methods of crushing ore today, as it has proven very satisfactory.

In addition to the above mills, there have been introduced in North Carolina from time to time various types of rotary pulverizers and pan amalgamators, some of which are supposedly to have been improvements on the stamp mill. Some of the more important mills introduced are the *Howland mill,* a flat circular disc revolving in an iron shell; the *Crawford mill,* with revolving iron balls, and the *Huntington mill.* In addition to these, there were the *Parson mill,* not unlike the Howland, but covered with a hood, and having the interior grinding surfaces coated with lead-amalgam; the *Meech mill,* in which the quick silver was comminuted by superheated steam; the *Wiswell mill,* being practically an iron Chilean mill fed with corrosive sublimate in connection with the electric current; the *Nobles process,* in which the ore was ground to 100 mesh between buhr-stones and the pulp run over amalgamated slabs of zinc or lead. Revolving *Freiberg barrels* were also used at some of the mines. The *Blake system* of fine crushing, combined with subsequent wet grinding, was introduced in 1884 but was abandoned in favor of the present stamp mill.

The above list of mills is some of the examples of the vast number of mechanical appliances used in the mines of the southern states for grinding and amalgamation. Most of these mills have been abandoned and replaced by the *stamp battery.* The stamp battery has proved itself far superior to any of the above listed mills, as most of them clearly demonstrated that such grinding apparatus produces float gold and flours the quick silver. Then, too, the mechanism of the mills was subjected to greater wear and strain as compared with the stamp battery and plate amalgamation.

SULPHIDE ORES

In a great many of the mines, as soon as the water level was reached, sulphide ores were encountered. Most of the free-milling ores near the surface were exhausted; therefore other methods had to be employed to recover the gold from the sulphides.

As far back as 1852, roasting processes were introduced at some of the mines in North Carolina. The first of which we have a record in North Carolina was that introduced by Dr. Holland, of Massachusetts. He introduced a roasting process in which the sulphide concentrates were mixed with nitrate of potash or soda and roasted in a reverberatory furnace at a low heat.

In 1856, a process for roasting sulphides, with subsequent amalgamation, was introduced by C. Ringel, at a mine near Rutherfordton, probably the Alta Mine. This process was afterwards practiced at the Gold Hill and other mines on old tailings with varying success.

In 1871-2, there was introduced at the Silver Hill Mine a method of making white lead-zinc oxide by the *Bartlett process*. This method consisted in roasting the galena-blende concentrates and condensing the zinc-lead oxide fumes. The material thus produced was used in the manufacture of paint. It is reported that this process was carried on successfully until all of the available suitable material was exhausted.

CHEMICAL TREATMENT

The first successful chemical treatment of sulphides was accomplished in 1879 by the introduction of the *chlorination process*. The plant was erected at the Phoenix Mine under the management of Mr. Adolph Thies. The process was known as the *Mears process,* and was improved upon by Mr. Thies, and was later known as the *Thies process*.

In 1880, a chlorination plant, employing the Davis process, was erected two miles south of Salisbury. This process differed from the Mears only in the method of precipitating the gold with charcoal instead of ferrous sulphate. Another Davis plant was erected at the Reimer Mine, in 1881, but soon after completion the plant burned down and was not thoroughly tested.

The cyanide process was not attempted in North Carolina for the recovery of the gold from ores until May, 1892. At this time Mr. Richard Eames, Salisbury, North Carolina, did considerable experimenting with the cyanide mill at the Gold Hill Mine. He succeeded in extracting only about 60 per cent of the assay value. At this mine it may be said that the cyanide process was not successful.

In the summer of 1893, a 10-ton cyanide mill was in operation at the Moratock Mine in Montgomery County. However, this plant was soon abandoned on account of the low-grade and the character of the ore.

In 1895, a cyanide plant was built at the Sawyer Mine in Randolph County for experimental purposes primarily. This plant was also soon abandoned. Again, in 1896, a 30-ton cyanide plant was erected at the Russell Mine, and during the same year a small plant was also built at the Burns (Cabin Creek) Mine, Moore County. Little is known as to the success or failure of these two mills.

OTHER CHEMICAL PROCESSES*

The *Hunt* and *Douglas process* was successfully applied in 1880 to the ores of the Conrad Hill Mine. The roasted sulphurets were leached with a ferrous chloride solution, converting the copper to a soluble chloride, from which it was precipitated as metallic cement on scrap iron.

The Designolle process, which consisted in treating the roasted ore with corrosive sublimate in iron vessels, was only

*Bulletin No. 10, 1897.

moderately successful in its application, for the reason that it made a very base bullion, the iron of the apparatus invariably precipitating any soluble salts formed in the roasting. It was worked for a time, during 1882-83, at a custom plant near Charlotte; at the New Discovery Mine, Rowan County, (1883); and at the Haile Mine, S. C., (1883).

A plant for the extraction of gold from pyritic concentrates, with the recovery of the sulphuric acid, was erected early in the decade (1890-1900) at Blacksburg, S. C., mainly for the treatment of custom ores. The concentrates were roasted in a Walker-Carter muffle furnace, which was connected with lead chambers. The amalgamation of the roasted product was carried on by a patent process known as the *Caloric Reduction Company's process,* the principle of which was a volatization of mercury into the mass of the pulp, followed by a condensation of the same, the amalgam being led into settling vats. It was proposed to use the tailing residues for the manufacture of red paint. The scheme, as might have been predicted, was a failure. A similar process, known as the *Phelps process,* had already been unsuccessfully tried on North Carolina ores, in (about) 1877, in an experimental plant situated at Philadelphia.

Attempts at pyritic smelting were made as early as 1847 at the Vaucluse Mine in Virginia by Commodore Stockton, but resulted in failure.

Matte smelting, followed by refining in reverberatory furnaces, was practiced (about 1881-1882) on the copper ores of the Conrad Hill and the North State mines in North Carolina.

Experiments on matting auriferous sulphurets from the Haile Mine, S. C., were made in 1886 by Mr. E. G. Spilsbury, but proved unsuccessful.

Regarding smelting processes in the South, probably most has been done in the attempted treatment of the complex galena-blende ores, carrying silver and gold, of the Silver Hill and Silver Valley mines, Davidson County, N. C.

The process in use at Silver Hill as early as 1853 was head-roasting, followed by wet-crushing in a stamp battery, the zinc oxide being dissolved and recovered separately, after which the residues were smelted in the old-fashioned Scotch open-hearth lead furnace, and the precious metals were recovered from the pig lead by refining in a cupellation furnace.

During the past years a number of patent processes have been experimentally tried on the Silver Valley ores in a plant situated at Thomasville, N. C., but it was not until 1895 that a successful process was introduced by Mr. Nininger, of Newark, N. J. It consisted of a down-draught jacket furnace, through which the fumes of lead and zinc were carried downward into condensers, where they were met by a spray of water, the liquor being led to vats where the lead oxide was deposited, while the zinc remained in solution and was subsequently precipitated as zinc oxide. The matte, carrying copper, gold, and most of the silver, was tapped from the well of the furnace and cast into pigs.

RECENT PROCESSES

Since the revival of gold mining in North Carolina in 1930, two other processes were experimented with, and both were soon abandoned.

The first process was tried on the saprolite ores of the Portis Mine in Nash County and later at the Black Ankle Mine in Montgomery County. It is reported that over $150,000 were spent in trying to prove this particular process, which was known as the *centrifugal* or *Lewis process.* The soft decomposed rock or saprolite was dug by a steam shovel and delivered to the plant by dump trucks. The ore was dumped on a grizzly, the large fragments of quartz and hard rock were thrown aside, and the fines dropped on bucket elevator. The bucket elevator hoisted the material to the top of the plant and emptied it into a 40-ton log washer or disintegrator. From the log washer the material emptied on a screen. The course material was stacked for later grinding in ball mill, while the fines went direct by trough and pipe into the centrifugal machine.

The centrifugal machine consisted of cylindrical bowl mounted in such a manner that it would revolve at varying speeds. The centrifugal machines were used in batteries of four and were said to handle 50 tons each of ore per day. The inside of the bowl was lined with horizontal grooves about $\frac{1}{4}$-inch in depth. The sludge or disintegrated material was fed by 4-inch pipe down the center, and the material was discharged at the bottom of the bowl. The bowl was revolved at varying speeds, depending on the consistency of the sludge entering it. The material was discharged by centrifugal motion over the sides of the bowl, while the mercury, and any gold that it might have picked up, was caught in the grooves. Since the ore consisted of a great deal of very plastic clay which had the tendency to pack into the grooves when the machine revolved, thus causing some of the mercury and gold to be discharged over the rim, the process was a complete failure. The ore, according to assays, showed an average of about $2 per ton, while the machine recovered only 40 to 60 cents per ton.

After this failure at the Portis Mine, the entire plant was moved to the Black Ankle Mine in the northeastern part of Montgomery County. After several futile attempts to operate this plant profitably, it was finally abandoned. A great deal of the machinery is still at the property.

The second process recently attempted in North Carolina is an *electrical* or *Gardner process*. The plant erected at the Parker Mine, near New London in Stanly County, did not prove successful and was soon abandoned. Various reasons are given for the abandoning of this process at the Parker Mine, among which are: the clay or saprolite did not contain sufficient gold; the process failed to recover the gold; and there was some dispute between the man financing the proposition and the engineer in charge.

The plant consisted of a revolving drum or disintegrator, sluice boxes, the necessary screens, and the electrical amalgamator. The ore was mined with drag, emptied on a grizzly, the coarse material piled for later grinding, and the fines shoveled by hand into the disintegrator. Some large quartz pebbles were also placed into the disintegrator to help break up the plastic clay. The sludge from the disintegrator emptied on a screen which took out bits of wood, leaves, and other foreign material. The clay sludge emptied into sluice boxes with riffles for collecting the coarse gold. The fine gold was to be recovered by the electric amalgamator.

The electric amalgamator consisted of two copper plates, one above the other and about one inch apart. The plates were silvered, the top plate on the bottom and the bottom plate on the top, then mercury applied on both plates. The two plates were used as the electrodes for an alternating current. As the sludge passed between the plates the electric current caused the precipitation of the gold either on the top or the bottom plate, depending on the direction of the flow of the current.

Since the above electrical amalgamator did not prove successful on the ores in North Carolina, further developments were made, and at the present time two of such machines are being used at the recovery plant built near Gold Hill to recover the gold from the old Gold Hill tailings. The present amalgamator consists of a shaking table about 2½ to 3 feet in width and 6 feet in length. Mercury pools are placed at regular intervals across the table, about three or four to each table. Immediately above the mercury pools there are three vertical aluminum plates. The table is covered with rubber, as this proved to be the best material for it. As the sludge is passed over the table the electric current is turned on and the table is made to vibrate. The process did not prove successful, and, after an expenditure of $8,000, the plant was abandoned.

A pilot plant of this type has been erected at the James Laboratory, Newark, N. J., by Mr. Frank K. Gardner, New York City, in cooperation with U. S. James. Several batches of low-grade ores have been shipped from North Carolina to this plant, and it is reported that the values have been recovered up to 98 per cent of that contained in the ore. All types of ores have been tried, the hard ores ground and the softer ores disintegrated by various methods. However, this process did not prove successful on the North Carolina ores.

SUCCESSFUL PROCESSES ADAPTED TO ORES OF NORTH CAROLINA

MECHANICAL PROCESSES

In many of the placers and the upper decomposed or oxidized portions of the veins, individual prospectors and miners have often made day wages, or better, with hand rockers. Various types of rockers have been used in North Carolina. The earliest rockers used consisted of logs hollowed out, sometimes with the ends open and sometimes with the ends closed. The soft material is placed in the rocker and by hand power rocked back and forth, the gold being concentrated on the bottom. The lighter material is thrown off, and, after considerable material has been washed and the concentrates become rather high-grade, the clean-up is either by hand-panning or by mercury being placed in the rocker.

Since the rockers did not handle the ore fast enough, sluice boxes, long toms, and various other appliances were used to speed up the operation. These boxes vary from **8** to 12 inches in width, **6** to **8** inches in depth, and from 4 to 10 feet in length. Across the bottoms of the boxes narrow strips of wood and other material are nailed to form riffles. A stream of water is fed into the upper end of the box, and the soft ore is shoveled by hand and thrown into the box. At regular intervals clean-ups are made; these clean-ups are determined by the richness of the ore as well as by the type of material washed. The riffles collect the gold, which is later concentrated by hand-panning or in rockers.

At a great number of places in North Carolina, hydraulicking has been attempted; some of the operations have proved quite successful. The most important localities where hydraulicking has been attempted are in Nash, Montgomery, Burke, and McDowell counties. The localities where the hydraulic material had no slope, a hydraulic lift or elevator was employed to raise the material.

The material to be hydraulicked was washed down by means of water under high pressure discharged through nozzles or "giants." The material thus washed down flowed through sluice boxes and the gold recovered as described above.

In some localities where the clay was very plastic, tenacious, or so tough that it was not easily hydraulicked, a disintegrator, Snodgrass machine, or blind trommel was used to disintegrate the clay. The materials from these machines were discharged through sluice boxes with riffles to amalgamation plates. These plates consisted of copper, about 3 by 5 feet, silvered and covered with mercury. Any gold that was not caught in the sluice boxes was usually caught on the plates. In some localities the material discharged from the plates was sent over burlap or blankets to collect the fine gold. The burlap or blankets were washed at regular intervals in bins.

In localities, such as the Portis Mine, where there was

considerable gold in the quartz fragments a screen was placed at the discharge end of the disintegrators to collect the quartz. This quartz was later sent to a stamp mill for grinding.

The *stamp mill* or (pounding mill), as it was originally called, was probably a European innovation. Probably the first stamp mill used in America was in 1836 at the Tellurium Mine in Virginia. A year later a Frenchman erected a small stamp mill at the Haile Mine in South Carolina. The first regular battery of stamps erected in North Carolina was that of the California type erected at the Kings Mountain Mine just after the Civil War, about 1866. The type of stamp mill generally used in North Carolina is a modification of the California type and has been manufactured for many years by the Mecklenburg Iron Works of Charlotte.

The stamp mill as manufactured at Charlotte is described briefly as follows: It consists of a battery of five stamps, each stamp weighing 750 pounds, and is given a 5 to 7 inch drop, 90 drops per minute. The mortar is of medium width and depth. At the back of the screen there is a large opening through which the foreign clogging matter can be cleared. The inside plates can also be taken out through this opening without disturbing the screen. On the discharge side of the mortar a 40-mesh, brass wire screen is placed to prevent large particles of the ore from dropping on the amalgamation plates.

The stamps, however, may vary in weight from 450 to 1,750 pounds, depending on the nature of the ore to be crushed. The weight of a 1,750-pound stamp is distributed as follows: weight of stem, 635 pounds; tappet, 250 pounds; boss head, 650 pounds; shoe, 250 pounds.

The stamp mill now in operation at the Thompson Mine consists of a jaw crusher, bucket elevator, ore bins of 40-

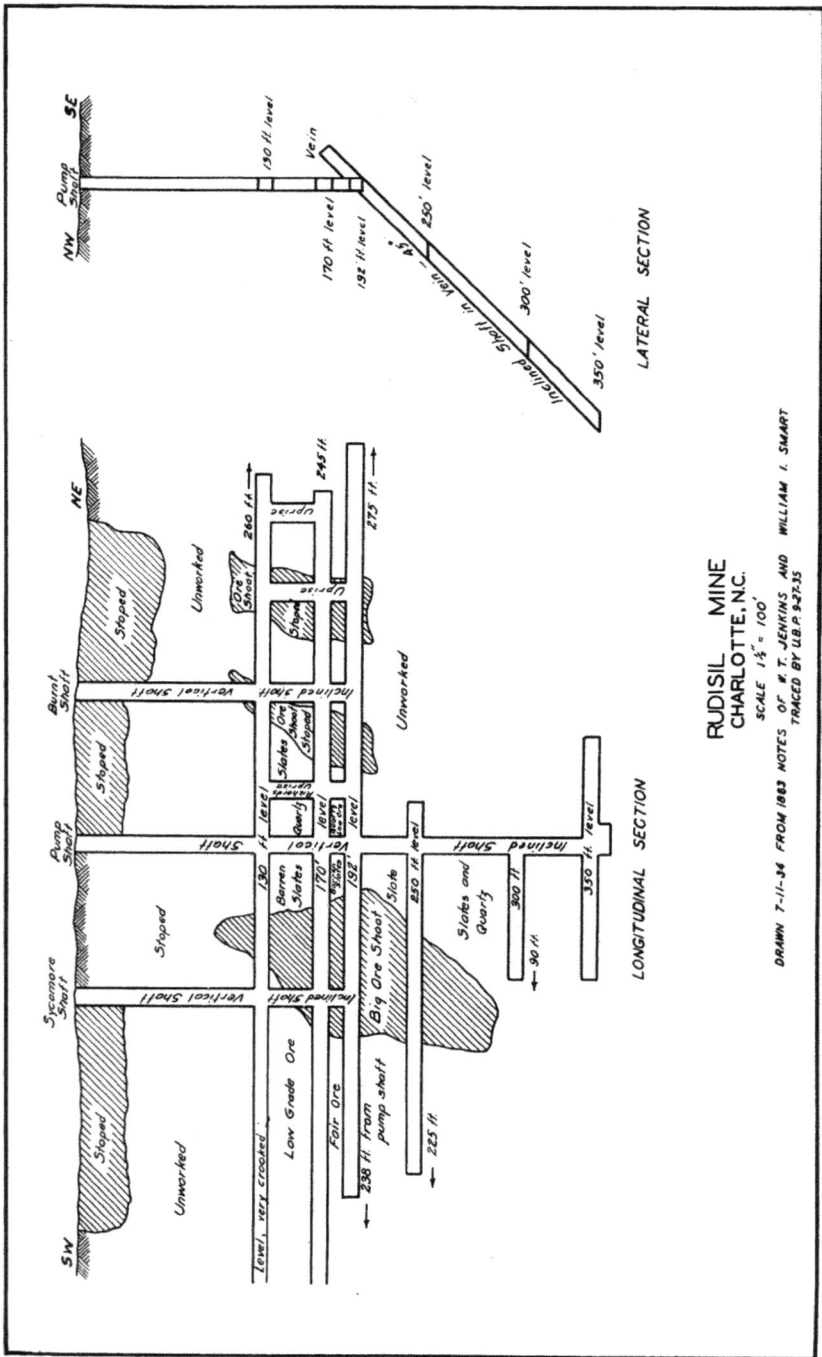

RUDISIL MINE
CHARLOTTE, N.C.
SCALE 1½" = 100'

DRAWN 7-11-34 FROM NOTES OF W. T. JENKINS AND WILLIAM I. SMART
TRACED BY U.B.P. 9-27-35

LATERAL SECTION

LONGITUDINAL SECTION

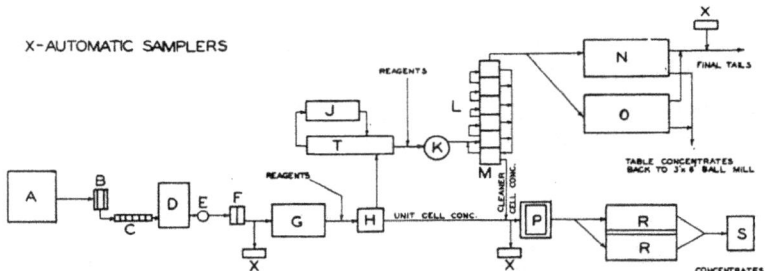

RUDISIL GOLD MINE INC.
FLOW SHEET OF CONCENTRATING MILL
DRAWN BY U. BIXBY PIERCE CH. ENG. SEPT. 27, 1935

X - AUTOMATIC SAMPLERS

REAGENTS

FINAL TAILS

TABLE CONCENTRATES
BACK TO 3'x 6' BALL MILL

UNIT CELL CONC.

CLEANER CELL CONC.

CONCENTRATES
TO SMELTER

A - 140 TON CRUDE ORE BIN
B - NO.3 JAW CRUSHER 12"x 18"
C - BUCKET CONVEYOR
D - 100 TON CRUSHED ORE BIN
E - AUTOMATIC FEEDER
F - ROLL-CRUSHER 12" x 18"
G - 4'x 6' COL. IRON WORKS BALL MILL
H - NO. 250 DENVER SUB-A UNIT CELL
T - DORR CLASSIFIER

J - 3'x 6' COL. IRON WORKS BALL MILL
K - DENVER SUB-A CONDITIONER
L - 5-NO.12 DENVER SUB-A ROUGHNER CELLS
M - 1-NO.12 DENVER SUB-A CLEANER CELL
N - PLATO-DEISTER TABLE
O - OVERSTROM TABLE
P - 15 TON PORTLAND ROTARY FILTER
R - CONCENTRATATE DRYING TRAYS
S - 40 TON CONCENTRATE BIN

ton capacity, two batteries of five stamps each, two amalgamation plates, and two wilfley tables. Each stamp will crush about one ton per twelve-hour day.

FLOTATION MILL

The flotation mill erected by the Rudisil Gold Mine, Inc., of Charlotte, is briefly described below:

Figure 3 gives the flow sheet in diagram.

The ore from the mine is delivered by truck to the flotation mill. The ore is dumped direct into 140-ton crude ore bin. From the crude ore bin the ore dumps direct into No. 3 jaw crusher, 12x18 inches. The crushed material is discharged into bucket elevator, which in turn discharges into 100-ton crushed ore bin. By automatic feeder the crushed ore is fed into a roll crusher, 12x18 inches. The ore from the roll crusher is fed into a 4x6 feet Colorado Iron Works ball mill. Between the roll crusher and the ball mill is an automatic sampler, from which samples are taken at certain intervals to keep a careful check on the character of the ore. The ore from the ball mill passes into a No. 250 Denver Sub-A unit cell. Between the ball mill and the unit cell the necessary reagents are added. The unit cell concentrates pass direct to a 50-ton Portland rotary filter, while the fines pass to a Dorr classifier. In closed circuit with the classifier is a 3x6 feet Colorado Iron Works ball mill. After the material leaves the classifier additional reagents are added. The material is then fed into a Denver Sub-A conditioner. The material from the conditioner passes to five No. 12 Denver Sub-A roughener cells. The cleaner cell concentrates are fed back to the unit cell concentrates and pass directly into the 15-ton Portland rotary filter. The remainder of the material from the roughener cells goes direct to the Plato-Deister, and overstrom tables. The table concentrates are fed back to the 3x6 ball mill, while the final tails are discharged into the tailings pond.

The discharge from the rotary filter passes to the concentrate drying trays. After the excess moisture has been driven off, the concentrates discharge into a 40-ton concentrate bin. The concentrates are shipped to smelter for treatment.

CHEMICAL PROCESSES

In 1880 the chlorination process was first introduced into North Carolina, at the Phoenix Mine, 7 miles southeast of Concord. The process adapted to these ores was known as the *Mears process,* but later developed by the manager, Captain A. Thies, into the *Thies process.* Below is a description of this process as adapted to North Carolina ores:

*Chlorination Process.**—The concentrates are hauled on the mine-railway to the chlorination plant. They are roasted in two double-hearth reverberatory and one revolving pan-furnace, the sulphur being reduced from about 43 to as low as $\frac{1}{4}$ per cent., and the value of the material being increased by 1/3. Each double-hearth furnace is worked by two men to a shift of 12 hours, the output being 2 tons of roasted concentrates for 24 hours for each furnace. The revolving pan-furnace is worked by three men per 24 hours, with the same output as the double-hearth. The fumes from these furnaces carry off into the air the equivalent of 13 tons of 50 per cent. sulphuric acid. The management has investigated the erection of lead chambers, but so far have not considered such an installation to their advantage. The Spence furnace has been tried at the Haile, without success. The roasted ore after cooling is elevated to the top floor of the chlorination house, 32 feet high. This consists of a four-story frame building, containing 3 chlorination-barrels, 11 filtering-tanks, 2 storage-tanks, and 13 precipitating vats. The ore is charged through a hopper into the chlorination-barrels by cars holding 1 ton each. The barrel is 60 inches

*Bulletin No. 10, "Gold Mining in North Carolina," 1897.

long by 42 inches in diameter, made of cast-iron and lead-lined (12 pounds of lead to the square foot). It also contains a lead value in order to ascertain whether the necessary amount of free chlorine is present. (The use of this valve is unnecessary after the character of the ores become known.)

The full charge consists of 120 gallons of water (to make an easily flowing pulp), from 8 to 11 pounds of bleaching powder, then the ore, and finally 12 to 15 pounds of sulphuric acid. The barrel is hermetically closed and revolves for about 3 hours at the rate of 15 to 18 revolutions per minute. (A 5 horse-power engine performs this work and also the elevating of the ore.) The barrel is then inverted, opened and discharged through a lead-lined semicircle in the floor to a filter on the floor below. There are 4 lead-lined filters to each barrel, their sizes being 6 by 8 feet by 18 inches deep in front and 17 inches in back. The bottom is covered with mineraline tiles 12x12 inches by 1 inch thick, perforated and having ½-inch gutters underneath; on top of these is placed a rack of 1¼ wooden slats, 4 inches high and 8 inches apart; the first layer above the tiles consists of 4 inches of coarse quartz pebbles (⅛ to ½ inch size), and this is covered by from 1 to 2 inches of ordinary clean sand. Before emptying the contents of the barrel, the filter is flooded with water to the level of the top of the filter-bed to act as a cushion. Then the original solution is passed through, striking on a float to prevent breaking the filter-bed. The ore-pulp is washed twice with clean water; the first time enough is added to stand 4 inches above the surface of the pulp, and the second time the tank is entirely filled.

This amount is found sufficient to thoroughly remove all traces of chloride of gold from the pulp (tests are made with $FeSO_4$). The filtered solutions are stored in two stock-tanks on the second floor, and are drawn off from these into the precipitating-tanks as required. The latter are 8 feet in diameter and 3 feet high, made of wood, the interior coated with asphalt. They are provided with three outlets, the upper one 18 inches from the top, the middle one 1 inch above the bottom and the lowest one in the jamb. The gold is precipitated in the metallic state with an excess of fresh ferrous sulphate, made in a small lead-lined tank. In warmer weather 48 hours suffice for settling, and in colder weather from 3 to 4 days. The supernatant liquor is drawn off through the two upper outlets, opened one after the other (in order to prevent any stirring of the precipitates), and passed through a box filled with sawdust to catch any precipitate. The gold precipitate is drawn from the tanks through a jamb-opening into a small lead-lined settling tank 2 by 2 by 4 feet. After standing 24 hours the supernatant liquor is siphoned off, and the precipitate filtered on paper. This is dried and mixed with about half its weight of borax and soda in almost equal proportions. Should iron salts be present, a little quartz sand is added. It is melted in graphite crucibles and cast into ingots of about 990 fineness. The whole operation is so simple that the most ordinary laborer can acquire the mechanical knowledge in a day. The repairs are practically *nil*.

Cyanide Process.—The first successful cyanide process in North Carolina for the recovery of gold from the ores was that of the Iola Mine in Montgomery County. Below is a description of this process by Percy E. Barbour.*

The cyanide mill at the Iola Mine was a continuous operation for a period of approximately 4 years. During this time the company shipped between 40 and 50 pounds of gold bullion semi-monthly. It was considered a financial success.

The ore from the mine was hoisted and dropped from an inclined skip over a 1¾-inch grizzly at the top of the head-

*Economic Paper No. 34, "The Mining Industry in North Carolina," 1914.

frame. The oversize dropped to an undersize bin in the lower portion of the headframe. The oversize was washed on a pickling floor at the foot of the grizzly with buckets of water as the ore fell from the skip, therefore no delay was occasioned and the rock was cleaned so that about 6 tons of waste were picked out by hand. The ore was then crushed to 1¾ inches in a 15x9-inch Blake crusher. The crushed ore dropped into the bin with the oversize through the grizzly, from which bin it was fed to a 18-inch belt conveyor. The belt conveyor was 80 feet long on centers, and inclined 22°. It was also the picking belt. At the lower end, a fine water spray cleaned the ore again and enabled the pickers, seated along the belt, to pick out about 6 tons more of the slate. The elevator discharged into a 100-ton flat-bottom bin.

There were two kinds of ore treated, which required radically different treatment of themselves, but could not have been mined and treated separately under local conditions, and therefore it had to be mixed just as it came from the stopes. The ore from the upper levels was oxidized and the accompanying waste was a soft clayey brown slate. In the lower levels the ore was very tough, hard gray quartz with a pearly luster and the country rock to which the vein was frozen was a tough gray slate. The unavoidable mixing of the ore as it happened to come out of the mine complicated the treatment problem, and more, also, the purely mechanical manipulation of the machinery adjustments.

After the ore was thoroughly cleaned it was sent to a stamp battery of five 1,750-pound stamps which had a drop of 7 inches and made 107 drops per minute. The battery pulp was elevated by a 12-inch bucket elevator to a standard double-rake Dorr classifier. The strength of the mill solution averaged 1.3 pounds KCN per ton and the protective alkalinity was 0.6 pounds per ton in terms of CaO. Owing to the hard and tough character of this ore, the slime was always granular, even when it passed 200 mesh (it was never colloidal), and it settled quickly and compactly. Therefore it was necessary, in order to obtain a clean sand for the tube-mill feed, to supplement the spray pipes in the bottom of the classifier with an additional pipe put in above the rakes. The Dorr classifier overflow went to a Dorr thickener and the classifier discharge dropped directly into the feed hopper of a spirally fed 4x20-foot tube mill.

The tube mill product was returned to the Dorr classifier by a 6x48-inch Frenier sand pump operating against a vertical head of 12 feet.

The overflow from the classifier went to a steel tank 18 feet in diameter by 8 feet deep, equipped with a Dorr continuous thickener. The thickener was not allowed to discharge continuously but drawn off at intervals of 30 minutes. Then only the thick pulp was withdrawn, fresh circulating solution was added to bring the specific gravity down to about 1.38, the most satisfactory consistence of pulp for agitation. The pulp discharged through a 12-mesh screen-bottomed tray into a boot of a second bucket elevator. This screen caught fragments of wood that the pickers failed to take out. The second elevator, similar to the first, carried the pulp to the top of No. 1 Parral tank. At the foot of the elevator additional lime was added in the form of an emulsion at the rate of 0.5 to 2 pounds per ton, depending upon whether the ore came from the bottom of the mine or the softer oxidized workings near the top. The pulp was discharged into the first Parral tank through a cone, made from an old cone classifier which had a pipe extending 5 feet below the surface of the contents of the tank. This was done to make the fresh pulp go into the cycle and to prevent any chance of short-circuiting. Fresh cyanide was introduced into the system at this point, being placed on a screen in the bottom of the cone, where it dissolved by the pulp splashing over it. The cyanide was added every 4 hours at the rate of 0.5 pound per ton of dry slime.

There were 3 steel Parral tanks, each 13 feet in diameter by 29 feet high with a capacity of 55 tons of dry slime in pulp of the average consistency. The first two tanks were arranged for continuous agitation, and the pulp, therefore, got an agitation treatment of more than 48 hours before it went into the third tank, which was originally a Parral, but which was changed. The air-lift pipe in this third tank was cut off one-third the way up from the bottom because the tank acted as a storage tank for the slime filter, which ran only during the day shift, being full in the morning and below the top of the air-lift pipe at night. The pulp, therefore, got further agitation in this tank, which made the total time of agitation equivalent to 52 hours. However, 90 per cent of the total extraction took place in the first tank. These Parral tanks were the usual type with 2 air-lifts near the circumference of the tank with their discharge pointed away from the diameter. These tanks each had a false bottom shaped like a peaked roof, made of wood. They were arranged so that either could be cut out and the contents filtered. The third Parral tank, which served the dual purpose of an agitation and storage tank, went through an interesting change from its original form. In as much as the filter was operated only during the day shift, this storage tank was during the night accumulating the slimes crushed during that shift and slowly filled; the reverse took place during the day when the tank was slowly emptied. Originally the tank was equipped with 2 air-lift pipes, similar to those in the other two tanks, except that the top tee was only about one-third the way up the tank. The discharge pipe to the filter was in the center of the bottom of the tank, so that no false bottom could be put in. The result was that the sand packed around this discharge had caused much trouble in keeping the pipe line to the filter open. To obviate this, one of the lift pipes was put over the center of the pipe and the other lift pipe disconnected, with the results which justified making this arrangement permanent. The slime naturally packed high along the circumference of the tank, and, sloping downward toward the center, built up a nearly perfect inverted conical bottom, so that to all intents and purposes it operated as a Pachuca tank.

The separation of the slime and pregnant solution was effected in a Kelly pressure filter, size No. 1B, which easily handled the total tonnage in less than 12 hours. The cycle of operation was: Filling press, 2 minutes; forming cake, 3 to 6 minutes, depending upon condition of leaves; returning excess pulp, 2 minutes; filling with wash pulp, 2 minutes; filling with wash water, 2 minutes; washing, 1½ to 3 minutes; returning wash and drying, 3 minutes; discharging, 12 minutes; total length of cycle, 25 to 30 minutes.

The filter treated 2.2 tons per charge. The dissolved gold in the tailing varied from 2 to 15 cents, an average of about 7 cents per ton. A solution wash was tried, but the additional time and labor required did not justify the slight additional saving of dissolved gold so that water was the only wash used. The tailing discharged from the filter leaves contained about 20 per cent moisture.

Built, as found necessary with this plant, on almost level ground, the problem of disposing of the tailing gave some trouble. There was a fair supply of water available, but not enough to carry the tailing through the long wooden launders with only a slight grade until the tough cake was completely broken up. At first this breaking up was accomplished by a real "down South hoeing nigger," who worked in the hopper with a hoe, but this was an unsatisfactory combination. A cake disintegrator was devised consisting of an oak log stuck full of wooden pegs arranged at right angles to the flow of the tailing and revolved 70 times per minute by belt and pulley. This disintegrator was at the very mouth of the discharge hopper and revolved so that the pegs tended to lift the cake up and carry it over the top of the log. The disintegrator was housed in to prevent splashing. The total water used in flushing away this tailing was about 2 tons per ton of dry slime.

Every 20 to 25 days the filter leaves were washed in a 0.5 per cent HCL solution. They were not treated in the filter drum but put in a vat and the solution forced through them in the same direction as the water, which forced the cake off the leaves. This treatment continued 24 hours.

The pulp and wash water were handled wholly by montejus; no pumps with moving parts were employed in connection with the filter. As before stated, when ground to 200 mesh, the product was all distinctly granular, and this product not only did not make a good cake despite all efforts, but owing to the rapid settling of the granular slimes, the tank in which the drum revolved was built up with this settled slime, so that every week it was necessary to shut down and sluice out this tank and lose the entire contents. Several air jets for agitating the pulp in the tank were tried without overcoming the difficulty.

The pregnant solution from the filtered slime went to a gold tank, from which it was pumped to a clarifying tank, which had a filter bottom made of wooden slats supporting cocoa matting, the space between the slats being filled with gravel containing pebbles not over one-half inch in diameter. On the cocoa matting were 3 inches of fine sand. The top of the sand layer had to be cleaned off every two weeks and the entire layer renewed every 2 months. The solution was handed to this tank by a 4-inch centrifugal pump.

From the gold-filter tank the solution flowed by gravity to three 6-compartment zinc boxes, 24x36x20 inches. Only four compartments of the six were packed. The zinc shavings, cut at the mill in an ordinary zinc lathe, graded 250 per inch and made up into skeins 2 feet long and as large around as a man's arm. These skeins were laid in the boxes with alternate layers crossing each other cobweb fashion.

Clean-ups were made twice per month, during which the small quantity of short zinc was treated with sulphuric acid; this procedure resulted in a cleaner bullion by excluding a portion of the zinc. The precipitate which passed a 30-mesh screen went to a clean-up filter press. The acid-treated material was put through the filter separately. The entire precipitate was then mixed, dried, and melted down in an iron-muffle furnace, which used coke for fuel. The flux used consisted of 25 parts borax, 15 soda, 15 sand, and 7 each of niter and slag. The slag from the melting was crushed, hand-jigged, and shipped to the smelter. The bullion assayed 718 fine ounces in gold and 225 in silver.

The total labor costs per day were $14 divided as follows: 1 mill man $2.50, 1 solution man $1.85, 1 mill helper $1.65, and 1 filter man at $2.

As shown by the above number of processes tried out in North Carolina, the State has been a sort of proving ground for processes. In a great many instances in North Carolina the owners of the processes have attempted to prove the process rather than to develop a mine.

In addition to the State being a proving ground, many of the property owners have asked exorbitant prices for the properties; hence, many properties that look promising have not been developed on account of the excessive prices.

Many of the plants which have been constructed have been mismanaged, and in many cases the manager has been inexperienced. As a result many gold mining propositions in the State have been complete failures when they should have been successful. Even today many people believe, even though they are without experience or training, that they can operate gold mines successfully. In some cases they invest their money, while in others they accept money from their friends for the development and operation of the properties. Gold mining today requires the best technical training and skill, and unless those in charge have these qualifications, gold mining to them is a dismal failure.

In some instances promoters have come into the State, opened up a small property, promoted the sale of stock and the construction of a mill before the ore body had been proven. As a result there are many failures, and in every case it gives a "black eye" to the gold mining industry in North Carolina. Many times mills are erected before the proper method for the treatment of the particular ore has been determined; therefore, the particular mill is often found unsuccessful for the ore and is soon abandoned. Within the past five-year period hundreds of thousands of dollars have been spent on propositions of this type.

It seems that the advisable thing to do in attempting to develop gold properties in North Carolina is to have reports on the properties by responsible engineers. These reports should not only give information as to the extent and value of the ore body, but should also give the best methods of treatment for the particular ore. In determining the extent and quality of the ore this can only be done through considerable exploratory work, as cross-cutting, shafting, and drift work, with subsequent careful sampling and assaying. After this has been accomplished, and the ore bodies show quantity and ore of such a tenor as to warrant commercial exploitation, a number of tests should be made on carefully selected average samples of the ore to determine the best methods of recovery. If the above procedure is carried out, there should be considerable activity in the gold mining in North Carolina as long as the price of gold remains at $35, or more, per ounce.

PRODUCTION OF GOLD IN NORTH CAROLINA

The following table, compiled from the production reports of the United States Mint, the U. S. Bureau of Mines, the North Carolina Geological and Economic Survey, and the Division of Mineral Resources of the Department of Conservation and Development, gives the figures for the amount of gold produced in North Carolina from 1799 to 1934, inclusive. These figures do not include the gold coined by C. Bechtler and the gold shipped direct to England from mines in North Carolina operated by foreign companies. Probably the amount produced in North Carolina would reach a total of fifty to sixty million dollars. Receipts from these foreign companies have been found totaling as high as $250,000 per month from small mines in Davidson County. Therefore, the total amount of gold shipped direct to England alone must have amounted to several millions of dollars.

GOLD AND SILVER PRODUCTION IN NORTH CAROLINA
FROM 1799 TO 1934

YEAR	AMOUNT	YEAR	AMOUNT
1799-1879	$19,659,600	1907	$ 96,494
1880	95,000	1908	98,163
1881	115,000	1909	43,399
1882	215,000	1910	73,474
1883	170,000	1911	70,782
1884	160,500	1912	168,999
1885	155,000	1913	127,543
1886	178,000	1914	131,984
1887	230,000	1915	172,744
1888	139,500	1916	26,673
1889	148,878	1917	13,102
1890	126,257	1918	1,648
1891	101,465	1919	1,055
1892	91,231	1920	1,158
1893	70,925	1921	1,713
1894	47,049	1922	1,948
1895	54,720	1923	1,166
1896	44,946	1924	4,561
1897	34,988	1925	18,615
1898	84,905	1926	1,644
1899	34,988	1927	1,018
1900	60,639	1928	13,515
1901	94,433	1929	5,054
1902	123,862	1930	32,963
1903	130,511	1931	12,956
1904	143,057	1932	10,423
1905	149,369	1933	18,522
1906	152,952	1934	41,835
		Total	$24,005,826

CHARACTER AND FINENESS OF GOLD IN NORTH CAROLINA

It is believed that the gold in North Carolina exists in the metallic state, with the exception of small quantities of telluride of gold at the Kings Mountain Mine, but it is invariably alloyed with silver, in proportions varying from 25 to 550 of silver.

The fineness of the native gold in North Carolina varies throughout the State, and to some extent with the different formations in which it occurs. In the northeastern section at the Portis Mine the gold varies from 925 to 950, rarely reaching 985. The gold from mines in Moore County varies in fineness from 700 to 750. The gold from the Davis and Stewart group of mines varies from 450 to 550, the lowest grade gold found in the State. Other mines of Union County, as the Phifer and Howie, produce a gold varying in fineness from 725 to 800, rarely reaching 850. The gold from the mines in the southern portion of Cabarrus County, as the Phoenix and the Reed, varies from 900 to 925 in fineness. At the mines in Mecklenburg County, as the Rudisil, Capps, and St. Catherine, the fineness of the gold is from 900 to 925 also. The Gold Hill, Whitney and Isenhour group produces a gold varying from 850 to 900 in fineness. Mr. Scott, operator of the Parker Mine, Stanly County, states that the gold from this mine shows 910 in fineness, and Mr. Phillip Eames states that the gold produced from the Young property in Davidson County varies from 950 to 975 fine. The mines west of the Blue Ridge produce gold varying in fineness from 900 to 980.

The question arises as to the cause of the variance in the fineness of gold throughout the formations of the State. Some have advanced the theory that the complete oxidation and weathering have increased the fineness of gold. This is not true, however, due to the fact that the gold at depth from the Davis-Stewart mines shows a fineness of from 450 to 550, while the gold at depth from the Gold Hill group shows a fineness of from 850 to 900. On the other hand the placer gold from the Parker Mine shows a fineness of 910, while the gold from the deep mines in Mecklenburg and Cabarrus counties (Phoenix, Reed, Capps, Rudisil) shows a fineness up to 925. Also the mines in the western part of the State, some of which have been worked below the oxidized zone, show a fineness up to 980, or a fineness much greater than the placer gold from Montgomery and Moore counties. In these latter counties the gold is alloyed with from 250 to 300 parts silver, therefore reducing the fineness of the gold as shipped to the mint to as low as 700 fine.

In some sections the gold includes or contains small cavities which are filled especially with copper and iron. These metals are not alloyed with the gold as the silver in other sections, but merely fill the openings.

CHAPTER III

Geographical and Geological Distribution of Gold Ores in North Carolina

Rocks of Gold-Bearing Areas; Nature and Structure of Ore Deposits; Weathering of Ore Bodies; Origin of Ores; Age of Ores

GENERAL STATEMENT

Geographically the gold ores of North Carolina are confined to two chief physiological divisions, the Piedmont Section and the Mountain Region. So far no gold ores have been found in the coastal plain formations other than that along the streams in Halifax and Nash counties. In these counties the gold occurs as placer material in the fluviatile sands and gravels along the streams, which have been transported from the gold-bearing formations to the northwest.

The most important gold-bearing section of the State is the Piedmont region. The ores in this section outcrop just west of the coastal plain deposits, beginning at the Virginia line on the northeast and extending to the South Carolina line on the southwest. Gold-bearing formations are found in the upper Piedmont section, outcropping in Yadkin, Catawba, Lincoln, Caldwell, Burke, and Rutherford counties. The most important gold-bearing formations are confined to the southeastern section of the Piedmont region, but isolated outcrops are found distributed throughout the entire region. Sixty-four of the counties in the Piedmont and Mountain regions have shown gold production at various times during the past. The largest production, however, has been confined to Nash, Montgomery, Stanly, Union, Mecklenburg, Cabarrus, Rowan, Davidson, Burke, and Rutherford counties. Recent geologic investigations, prospecting, and development work have been confined to the above listed counties.

Geologically the gold ores of North Carolina may be divided into six chief groups as outlined in Bulletin No. 3, "Gold Deposits of North Carolina." These belts are outlined as follows:

1. The Eastern Carolina Belt.
2. The Carolina Slate Belt.
3. The Carolina Igneous Belt.
4. The Kings Mountain Belt.
5. The South Mountain Belt.
6. The mines west of the Blue Ridge, or the Western Belt.

These gold-bearing belts have no definite outline or limits, especially is this true between the Carolina Slate Belt and the Carolina Igneous Belt.

The above outline is given primarily for the purpose of simplifying the geologic description of each area. The geologic formations of each belt are more or less related but can be described separately.

ROCKS OF THE GOLD-BEARING AREAS

The rocks of the gold-bearing areas of North Carolina may be divided into two chief groups, igneous and metamorphic. The igneous rocks may be divided into two chief classes, acid and basic. The acid rocks consist largely of granites, those varying in color from the light gray to pink, rhyolite, the acid tuff and breccia. The basic group includes gabbro, diabase, diorite, andesite, and the basic tuff and breccia. The metamorphic rocks may be divided into gneisses, schists, and slates.

Among the acid rocks the granites are by far the more important. These rocks are confined largely to the Raleigh Granite Belt of Vance, Franklin, Granville, and Wake counties, and to the Main Granite Belt, which extends from

Caswell and Person counties on the north in a southwest direction across the counties of Alamance, Guilford, Davie, Davidson, Rowan, Cabarrus and Mecklenburg.

The rhyolite, acid tuff and breccia are confined largely to the area between the Raleigh Granite Belt and the coastal plain deposits on the east and the Main Granite Belt on the west; especially is this true in the counties of Moore, Montgomery, Randolph, Rowan, Cabarrus, and Union. The rhyolite usually occurs more or less lenticular, seldom more than a few hundred feet in width and less than a half mile in length. The tuff and breccia are found in very small lenticular deposits, seldom more than a hundred feet in width and rarely, if ever, more than two or three hundred feet in length. The breccia is usually confined to zones between the rhyolite and the country rock, as gneisses, schists, and slates.

The basic rocks usually occur as dikes and other small intrusions cutting the older rocks. These rocks are confined largely to the Carolina Slate Belt, but numerous outcrops also occur in the upper Piedmont section, as well as to a limited extent in the Mountain region. In the Piedmont section the more important topographic relief is due to intrusions of the basic rocks. These formations usually form the higher ridges in the lower Piedmont section.

The metamorphic rocks of the gold-bearing sections of the State are confined to the Carolina Slate Belt, the upper Piedmont section, and the Mountain Region. Locally the gneisses, schists, and slate are all known as slate. The gneisses are principally the granite gneisses with occasional lenticular mica gneiss zones, and may be of both sedimentary and igneous origin. The schists are largely mica schists, but numerous lenticular outcrops of pyrophyllitic (Moore and Montgomery counties), sericitic (Union, Randolph, northwest Montgomery counties), talcose, and chloritic. A very important group of the metamorphic rocks is the Monroe Slate Belt of the south-central part of the State.

The Monroe Slate Belt covers a greater part of Union and Stanly counties, the northwest section of Anson County, the Western half of Montgomery County, the southeastern parts of Davidson, Rowan, and Cabarrus counties.

The origin of the Monroe slate is not definitely known, but it is apparently a volcanic ash deposited both above and below water level, with lenses of land waste included in it. The outer limits of this slate area have been highly metamorphosed into sericitic, chloritic, and pyrophyllitic schists. In the circular area, especially in Union and Stanly counties, the slate is practically horizontal with occasional synclinal and anti-clinal structures. Into this slate are also intruded basic rocks, especially diabase and gabbro. The basic rocks are not confined to any definite areas but are more or less distributed throughout the entire slate belt.

The general strike of all the rocks of North Carolina is in a northeast-southwest direction. The dip of the formations varies considerably; in some localities they dip to the northwest, while in others to the southeast. The angle of the dip varies from practically horizontal to vertical.

NATURE AND STRUCTURE OF ORE DEPOSITS

The gold deposits in North Carolina may be classified as follows:

1. Veins.

2. Lodes or mineralized zones.

3. Placers.

4. Saprolite.

These deposits may be further divided as follows:

1. Veins.
   a. Fissure veins.
   b. Veins formed by growing crystals.

2. Lodes or mineralized zones.
    a. With quartz.
    b. Without quartz.

3. Placers.
    a. Stream gravels.
    b. Bench gravels.
    c. Residual gravels.
    d. Talus.

4. Saprolite.

VEINS

The gold ores of North Carolina which occur in veins are pretty well distributed throughout the Piedmont and Mountain sections. The fissures in which the veins occur were apparently openings in the rocks before the deposition of the ores. The vein material is usually quartz with such primary materials as pyrite, chalcopyrite, galena, sphalerite, argentite, and gold. The fissures, in which the vein material was deposited, were caused probably by dynamic forces from the northwest; these forces developed normal faulting with, as a rule, slight compression. Apparently many of the larger fissures thus caused were filled with basic dikes rather than with ore deposits. The principal gangue minerals associated with the above deposits are quartz, siderite (especially at the Phoenix Mine), and rhodochrosite (Snyder Mine). The secondary minerals are cerussite, chalcocite, malachite, tenorite, and anglesite. The principal secondary gangue minerals are limonite, hematite, chlorite, actinolite (Ward Mine), calcite (Phoenix Mine), sericite (especially at mines in Montgomery and Union counties).

The theory of the origin of the veins formed by growing crystals was first developed by George F. Becker and Arthur L. Day, in 1905.* They did considerable experimental work

---

*Becker, George F., and Day, Arthur L. The Linear Force of Growing Crystals. Wash. Acad, Sci., Proc., v. 7 (1905), pp. 283-288.

upon this subject with surprising results. They state: "It became recently certain that it (the force of growing crystals) is actually of the same order of magnitude as the ascertained resistance which the crystals offered to crushing stresses . . . . It is manifest that we here have to deal with a force of great geological importance. If quartz, during crystallization, exerts a pressure on the sides of a vein which is of the same order of magnitude as the resistance of the wall rocks, and it thus becomes possible that, as indicated by observation, the mother lode and other great veins have actually been widened to an important extent, perhaps as much as 100 per cent or even more, by pressure due to this cause."

This theory of the origin of the veins offers a plausible reason for such veins as are found at the Gold Hill, Parker, and similar mines.

LODES OR MINERALIZED ZONES

Another type of gold deposits in North Carolina, which has been worked as much or possibly more than the veins, is the lode or mineralized zone. This particular type of deposit may also be divided into two groups: (a) those with quartz stringers, (b) those with little or wholly without quartz. Mines of this type are especially the Russell, Black Ankle, Keystone, and Alford. Ore bodies of this type are usually very wide, sometimes a hundred feet or more, and of considerable length. However, it does not mean that the entire width of the zone is sufficiently mineralized and of such a grade as to be worked profitably across the entire zone, but there are streaks, 20 feet in width or more, that have been worked profitably. In some of the lode deposits there are numerous quartz stringers, which are apparently the source of the gold or have some connection with it; as in many localities where considerable panning has been done, the quartz stringers show the best panning. It is also impossible to determine by megascopic methods the lode or zone; therefore, it is necessary to pan the ore or have samples assayed at regular intervals.   In localities where there is little or no quartz in these mineralized zones, it is even harder to determine the width of the gold-bearing zone.   The gold occurs as thin leaves or flakes between the laminations of the schist and slate.   Due to the flaky nature, the gold is hard to recover.

In the upper zone or layer, the oxidized portion, the gold is usually free-milling and can be recovered by ordinary amalgamation; but as depth, especially to and below water level, is reached, the ore is refractory and the gold can not be recovered except by proper treatment, as roasting, chlorination, cyanidation, or flotation. Many of the old mines have been worked only to water level, or to a depth at which the sulphides are encountered, due to the cost of treatment of the ores.   Generally speaking, the wider the zone the lower the grade of ore.

The ores from the above types of deposits are usually very low-grade and seldom average more than two or three dollars per ton.

WEATHERING OF ORE BODIES

In North Carolina, as well as in other states in the southeastern district, the surface of the country rocks has been subjected to weathering agencies for a great many years. The rocks of this section therefore are weathered to a great depth, occasionally to 200 feet or more.   A mud slide occurred at a depth of 500 feet in one of the deeper mines of the State.   Also fissures filled with mud have been encountered in drilled wells at depths of 300 feet, or more, in the Piedmont section of the State.

The deep, weathered portion of the earth's surface in the southeastern section is due to the long period of weathering of the rocks, to the absence of glaciers which removed the weathered portion in sections further north, and to the physiographic features.

Many of the veins and lodes or mineralized zones have been weathered to a great depth, and free-milling ore is the result. The iron sulphides for the most part have been changed to brown hematite and limonite, the copper sulphides, as chalcopyrite, have become chrysocolla, rarely azurite and sometimes either the black or red copper oxide. In some mines, as the Fentress Mine in the southern part of Guilford County, the copper constituents have been almost entirely leached out. This is also true of the lead and zinc oxides in the Silver Hill and Silver Valley sections of Davidson County and, to a limited extent, of the lead-zinc sulphides in Montgomery County.

The decomposed brown ores in the upper zones contain free gold, sometimes in such quantities as to be easily visible to the naked eye. This is especially true at the Parker, Crowell, and Portis mines.

The method of working the upper or decomposed portions (gossan) has been entirely by mechanical methods, but as depth was reached, resulting in encountering complex and refractory ores, with a considerable decrease in value, more elaborate methods of recovery were necessary.

Above the water level, in the oxidized portions of the veins, there is also some enrichment due possibly to circulating manganiferous and alkaline waters, and possibly, to some extent, to weak organic acids.

According to the old reports, considerable profits were made from mining the oxidized ores, due primarily to the ease by which the gold was recovered as well as to the higher values in these upper zones. Also in the early days, due to the type of machinery used, the lack of efficient pumps, and the refractory properties of the ores, many of the mines were abandoned as soon as water level was reached. Even though many of these mines have been abandoned at water level, it does not follow that ores of such quantity and quality to be commercial will not be encountered at greater depth. Also, due to modern mining and milling methods, and to a greater knowledge of the recovery methods, many of these abandoned mines may prove to be of commercial value.

In some sections, as the Portis Mine in Nash County, the Parker Mine in Stanly County, and various sections in Montgomery County, due to the weathering of the ores and the resulting accumulation of free gold in the low places, very rich bodies of residual gravels have been encountered. Many of these have been worked at considerable profit as is evidenced by the amount of work conducted in these sections. The old miners and prospectors recovered the free gold and piled the gravel which contained values as if to mill it at a later date. These gravels at times are reported to average $5, or better, per ton.

The weathering of these upper zones and the resulting accumulation of this weathered material in valleys and along streams form the placer deposits throughout the gold-bearing sections.

ORIGIN OF GOLD ORES

In order to solve the problem of the origin of the gold ores in North Carolina, detailed underground studies of all the mines in the various sections of the State would have to be made. Since this has not been done, the origin of the ores has never been definitely settled. However, the gold ores were apparently deposited from solutions. The solutions probably carried large amounts of silica since the veins or zones in which the gold occurs are composed largely of silica, together with small quantities of the primary minerals, as pyrite, galena, sphalerite, chalcopyrite, and gold. In the Silver Valley section there were also considerable amounts of argentine and native silver.

The source of the solutions forming the gold deposits in North Carolina is not definitely known, but it was apparently from deep-seated granitic magmas. In some sections, however, there seems to be some relation or connection between the basic intrusions and the gold deposits. No definite conclusions can be drawn as to the source of the solutions until further studies are made.

The deposition of the ore occurred after the formation of the fissures in the rocks. The solutions coming from beneath, through the release of pressure and lowering of the temperature, deposited the ores in the fissures.

AGE OF ORES

The age of the gold ores in North Carolina has never been definitely established. All of the rocks in which the gold veins occur in the Piedmont section are of pre-Cambrian Age except those in the Kings Mountain Belt. The rocks of the Kings Mountain Belt have been put into the Cambrian group of early Palaeozoic Age.* Since the rocks of the Kings Mountain area contain gold deposits, the ores must be post-Cambrian. The Triassic formations on the east, especially the basal conglomerate, contain gold; therefore, the gold deposits must be pre-Triassic. The age of the gold deposits then may be Palaeozoic in age.

---

*Gaffney-Kings Mountain Folio by Arthur Keith, 1931.

CHAPTER IV

THE EASTERN CAROLINA BELT

GENERAL DESCRIPTION

The gold ores of the Eastern Carolina Belt are confined principally to an area covering approximately 300 square miles in the northern parts of Nash and Franklin counties and in the southern parts of Warren and Halifax counties, near the village of Wood. The topography of this section is gently rolling with elevations at no place greater than 100 feet above the stream levels. Due to the gently rolling nature of the topography, no deep gulches or gorges are found in this section.

The rocks of the area are principally gneisses and schists with intrusions of diorite and occasionally diabase. The country rock, composed of gneisses and schists, is weathered to a great depth, as at no place have shafts been sunk that encounter hard rock. However, where the diorite intrusions occur, shafts have encountered hard rock at very shallow depth. Also occasionally on the hillsides small knobs of the diabase intrusions are exposed. These intrusions are very narrow, seldom exceeding 3 or 4 feet in width.

The gold-bearing rocks of the Eastern Carolina Belt are bound on the west by the Louisburg granite and on the east they are overlain with coastal plain deposits. All of the rocks in this section are apparently pre-Cambrian in age.

The gneisses and schists are cut by numerous quartz veins usually stringers seldom exceeding 2 or 3 inches in width, but occasionally large masses of quartz are found. These bold exposures of quartz are due to quartz veins, always barren, which reach 15 to 18 feet in width. The gold is apparently confined to the stringer veins which intersect the rock in almost every direction. The most predominating trend, however, is in a northeast-southwest direction, paralleling the strike of the old formations.

The quartz veins on decomposing (literally) cover the surface of the ground over large areas and tend to accumulate more or less in the shallow valleys. Most of this accumulation is due to ordinary creep and slumping of the water-laden decomposed rock. However, along the streams, as Shocco and Fishing creeks, there are large areas underlain with more or less rounded gravels. These gravels are apparently transported due to their rounded nature. Some of these valley flats have been drilled with Empire drills to determine the value of the gravels for the purpose of dredge mining. Since the gravels seldom exceed 2 or 3 feet in thickness and usually have an overburden of 3 to 8 feet, little success was had when dredging was attempted. Some state, however, that the failure of the dredging was not due to the low value of the gravels but to litigation and the inability of the owners of the dredge to secure sufficient property.

Recent investigations along these lines on either side of Fishing Creek have revealed gravels of sufficient value to be worked profitably by dredging, according to information received from Mr. A. L. McNeer, placer miner of wide experience in the western fields. At the heads of many of the shallow valleys there are deposits of gravel, some with little overburden, others with considerable overburden, which may be worked profitably provided the properties could be had at reasonable price and the operations conducted under proper management.

The veins, especially the stringer veins, are composed of sugary or saccharoidal quartz, which lends itself to inexpensive crushing methods. At a number of localities, more especially along Gold Mine Branch, large piles of the quartz are found, which were accumulated during early hydraulicking operations. However, some of the piles of quartz have been due to the miners hauling the material by wheel-barrow, wagons, and other methods, from the hillsides to the stream and there washed.

A large amount of the quartz thus accumulated contains values which may be capable of being milled at a small profit. Copies of assays on the quartz, taken from the old reports, show values better than $3 per ton.   More recent sampling of a great many of these quartz piles has shown values from $2 to $14 per ton with an average of about $3 per ton with gold at $20.67 per ounce.   At the present price of $35 per ounce the average would be $4 to $5 per ton.*

MINES IN THE EASTERN CAROLINA BELT

NASH COUNTY

Several mines have been worked in Nash County, but the Portis Mine has been by far the most important.

Portis Mine.   The Portis Mine is situated just south of Ransom's bridge in the northeast corner of Franklin County about 18 miles east of Louisburg.   The highest point on the property is 108 feet above Shocco Creek from which water in the early days was taken for hydraulicking purposes.

The principal rock is schist, considerably decomposed and usually beyond recognition, into which have been intruded diorite and diabase.   The most important dioritic intrusions are in the form of sills one above the other and separated with only a few feet of country rock.   The upper diorite sill is about 9 feet in thickness, while the lower is 14 feet and possibly more.   These sills dip in a westerly direction at a low angle, as one of the miners stated that a hole drilled 300 yards west of the outcrop encountered the sill at a depth of about 40 feet.

The weathering of these diorite sills forms a very light colored clay, almost white at times, which is known locally

---

*Assay value for old mines $20.67; for new development work $35 per ounce.

Fig. 4. Entrance to Mine, Norlina Mining Company, Wood, N. C.

Fig. 5. Removing Overburden, Norlina Mining Company, Wood, N. C.

Fig. 6. Plant of Norlina Mining Company, Wood, N. C.

Fig. 7. Diesel Locomotive, Norlina Mining Company, Wood, N. C.

as the "White Belt." The White Belt is cut by numerous quartz veins, which criss-cross and have no definite trend. In a tunnel in the upper diorite sill, numerous quartz stringers, more or less reticulated, cut the diorite. These quartz veins are always accompanied with considerable manganese and iron oxide. At times these small veins show free gold, and almost always, when carefully ground and panned, show a great number of colors of both fine and coarse gold. The weathering of these diorite sills and the breaking down of the quartz veins causes an accumulation of the quartz and gold in the valleys. In some places this detrital material shows a depth of from 15 to 30 feet, with gold occurring from the top to the bottom but with a greater concentration on the bed rock.

The sills, when first intruded, extended a great deal farther toward the east than the present outcrop. The weathering of these sills, with the contained gold, was the source of a great deal of the placer gold found between the outcrop and Fishing Creek. However, in addition to the gold distributed by the weathering of the diorite sills, there are numerous quartz veins intersecting the country rock. Due to this condition a great deal of the surface material, as well as the decomposed rock, can probably be worked profitably. In an area immediately to the east of the White Belt an engineer drilled holes and carefully sampled quite a large area, which he stated averaged better than $2 per ton. Also, numerous samples which have been taken of the quartz covering the surface of the ground showed values of approximately $3 to $5 per ton.

On the east side of Gold Mine Branch, near the headwaters, is an ore body which has been worked considerably. This ore body consists of numerous quartz veins varying from stringers up to 8 or 10 inches in thickness, occasionally 2 feet, which are close enough together to work across the entire width. At this particular place the zone has been worked to a width of 50 feet or more, over 300 feet in length, and to a depth of approximately 30 feet. The country rock in this immediate vicinity is highly metamorphosed schist which also contains considerable gold. Almost every pan of the material panned shows a number of colors, at times in sufficient number to "string the pan."

In the spring of 1935 the Norlina Mining Company, Lansing, Michigan, acquired the Portis property along with the White House property which lies between the Portis and Fishing Creek. The total acreage controlled by the company is 1,668 acres, 955 in the Portis tract and 713 in the White House tract. Mr. W. L. Long, of Raleigh, secured the property from the Dolan estate, Philadelphia, Pennsylvania, and sold it to the Norlina Mining Company. Mr. R. E. Olds, of Detroit, is the President; Mr. C. S. Robertson, formerly of Nova Scotia, Vice-President and general manager; Mr. W. A. Milliken, assistant manager; and Mr. J. C. Lettellar, Detroit, Michigan, is the geologist and chemist.

Mr. Phil Sturgess, Wood, North Carolina, has long been enthusiastic about the possibilities of these properties. Probably no man knows the properties better than he. During the past 30 or 40 years, Mr. Sturgess has practically made a living from panning along the streams and from working the gravels in sluice boxes. The property was formerly owned by Mr. Sturgess' father, Colonel Sturgess.

Through the efforts of Mr. C. S. Robertson the property is to be developed. He spent considerable time and money in prospecting and developing the property to determine the value and extent of the ore bodies as well as to determine the best methods of recovering the gold from quartz, decomposed rock, and the clay. Apparently he is satisfied with the results of his explorations and tests, as approximately $150,000 have been expended in the erection of a recovery plant.

The ore on and near the surface will be mined with drag line and delivered to the plant by tram. Also, a shaft is being sunk on the quartz zone, and the material from this shaft will also be delivered to the plant by tram. The tram line will be approximately one-fourth mile in length. A gas locomotive will be used to transport the automatic dump cars of four-tons capacity each from the ore body to the plant.

The main building housing the machinery is 60x110 feet and 66 feet high. The plant is constructed of best material on concrete foundation and covered with galvanized sheet metal.

The ore is hoisted to the top of the plant and dumped automatically into a trommel. The trommel is 3x20 feet, the first 12 feet are blind and the last 8 feet equipped with one-fourth inch openings. The oversize from the trommel drops direct to the crusher. The crusher is a gyratory with a capacity of 1,000 tons per 24-hour day. The fines, one-fourth inch or less, go by gravity to 12 gigs. From the gigs the ore passes to an Aiken classifier, 54 inches by 16 feet. The classifier also acts as a dewaterer. The material drops on bucket elevator which carries the ore to the ore bin, 480-tons capacity. The crusher, gigs, and elevator are powered by 180 H. P. Anderson Diesel engine.

The ore from the bins drops direct into stamp batteries, ten batteries of five stamps, each weighing 950 pounds. From the mortars the crushed material, minus 24 mesh, passes to the amalgamation plates. There are two plates, 4.6x8, for amalgamation. The plates are in two pieces with one inch drop between. As the crushed material drops on the plates, additional water is added to thin the sludge. The first plate is equipped with an electric baffle, which is said to activate the mercury to keep it from flouring and to cause a precipitation of the flour gold.

From the plates the material passes over English blankets, one blanket to each set of plates, 4.6x4 feet. Any mercury escaping from the plates is caught in mercury trap, each set of plates being equipped with a trap. The material from the plates and first set of blankets passes by trough to the second set of blankets covering 1,500 square feet. The blankets are washed at regular intervals in a bin. Any fine gold that is caught on the blankets is emptied into a bin and is recovered in an amalgamating barrel. All retorting is done by the company in the assay laboratory.

The stamp batteries are powered by 130 H. P. Franklin Diesel. A 100 H. P. Hill Diesel engine is on hand for mill when needed.

Electric power will be provided by direct connected generator, 12.5 K. W., to the 130 H. P. Diesel.

The shaft will be a three-compartment, equipped with 529-foot belt-driven compressor powered by 100 H. P. Hill Diesel engine. The hoist will be powered by 130 H. P. Diesel. Electric power for lights in the shaft and drifts will be supplied by 30 K. W. direct connected generator.

The water supply consists of a six million gallon reservoir located in the valley just below the plant. The water will be pumped from the reservoir to a 5,000-gallon tank at the top of the mill. Water from this tank will be fed direct to stamps and gigs.

A private six-room residence has been erected for the superintendent. A bunk house of fourteen-men capacity has also been erected. In addition, there is an office, dining room for 20 men, a complete blacksmith and machine shop. The assay laboratory is fully equipped with small testing plant and all materials and equipment for assaying.

The great problem to overcome in recovering gold in this section is to disintegrate or digest the clay economically.

Several processes have been attempted at this mine, but so far all have been complete failures. Mr. Robertson stated that he was more than pleased with the quantity and quality of the ore in sight and is fully convinced that the clay problem has been solved. If he has succeeded in solving the problem, there is little doubt as to the future financial success of the company.

HISTORY OF PORTIS MINE

From the old records discussing the Portis Mine, apparently the first authentic discovery of gold in this section dates back to 1835, or approximately 100 years ago. Soon after the discovery of the gold at this mine, a real mining camp was built. A great number of stories are told locally about the early operations, especially relative to the number of people employed, the personnel and activities of the companies. The stories related are quite similar to the history of the western mining camps. In the boom days hundreds of people were employed; whiskey and rum flowed freely resulting in a number of murders. The small graveyards nearby are evidence of these facts.

Even in the later 80's a large number of people were employed as tributors in working the gravels in the valleys and sluicing the decomposed formations on the hillsides. At present, development will probably show even greater activity than in former days. The chief difference, however, in the present development is the difference in the type of personnel operating the properties and the modern up-to-date methods of mining and milling the ore.

As to the values of the ores in the early days, Colonel Sturgess states "that 1,000 cubic yards yielded 1,018 pennyweight of gold, the loose vein rock obtained averaged about $8 per ton, assay value." The amount of gold recovered, when based on the present gold value from the material suitable for hydraulicking, would average close to $2 per yard and the quartz between ten and fifteen dollars per ton. These values correspond very closely to the values obtained by investigations of the Division of Mineral Resources and the present company developing the property.

The *Mann-Arrington Mine* is located in the northwest corner of Nash County, near Argo, five miles to the southeast of Ransom's bridge. In the vicinity of this mine the country rock is a chloritic schist, at times porphyritic, with some of the phenocrysts one-eighth inch in diameter. The country rock may be a highly metamorphosed diorite.

The general strike of the formations is N. 58° E., and the dip 40° toward the southeast. At depth the rock contains iron sulphides and quartz lenses from stringers up to 12 inches or more in thickness. The quartz veins are usually interlaminated in the schist, but occasionally they are found cutting the schistosity at low angles. The quartz, like that at the Portis Mine, is quite saccharoidal and often of a reddish brown color due to the oxidation of the sulphides. The quartz also often contains included fragments of the chlorite. The deepest shaft so far known was worked to a depth of 108 feet. The last real work done at this property was in 1894. However, within the past five years some exploratory work was done, but apparently this did not prove a deposit of value, as it was soon abandoned.

The *Arrington Mine* is in Nash County, one mile southeast of the Portis. Next to the Portis Mine this is the best known mine in the region.

The *Conyer's Mine* is seven miles from Whitakers, on Fishing Creek. A shaft 30 feet in depth was sunk on an 18-inch vein which is said to have shown milling ore all the way. The vein was quartz carrying brown hematite and sulphide ores. In addition to this vein, considerable placer material was also worked.

Other mines in this section which have shown some production, but less prominent than the mines mentioned above, are the Nick Arrington, Thomas, Kerney, Taylor, Mann, and Davis. East of this gold-bearing section some gravels were worked, but very little systematic mining was carried on.

The production of gold in the Eastern Carolina Belt is estimated between two and four million dollars. From the amount of hydraulicking done, covering several hundred acres, the latter figure is nearer correct. If the placer material averaged $2, or better, per ton, as is indicated by recent sampling, and several million tons of material have been hydraulicked, the production must have been close to four million dollars.

CHAPTER V

GENERAL STATEMENT

The Carolina Slate Belt is the largest and most important gold-bearing section in North Carolina. The term slate used in this connection covers a broad designation. It consists of argillaceous, sericitic, and chloritic metamorphosed slates, crystalline schists, and true slates. In addition to these rocks, there are numerous lenticular intrusions of volcanics, such as rhyolites, porphyries, breccias, and basic intrusions, especially diabase. In this whole area there are also intensely sheared phases of the above rocks.

This slate belt begins in Person County on the north, where it is about fifteen miles in width, and extends in a southwest direction covering part or all of the counties of Person, Durham, Orange, Alamance, Chatham, Randolph, Moore, Montgomery, Davidson, Stanly, Cabarrus, Anson, and Union, and passes into the northern part of South Carolina. The greatest width is approximately fifty miles. The slate belt is bounded on the west by the Central Granite Belt and on the east by the Henderson granite area and the Triassic sandstone belt. This slate belt is the so-called "great slate belt" of Olmsted, the "Taconic slates" of Emmons, and the "Huronian slates" of Kerr. Most of the rocks in this entire belt are volcanic in origin, but occasionally rather broad lenses of old land waste are included in them.

In the south-central part of the State a large area of this slate belt has been given the name "Monroe slates." In the central part of this area the slates are generally more or less horizontal in position, while near the granitic intrusions on the east and west they stand almost vertically and have been metamorphosed into pyrophyllitic (talcose), sericitic, and chloritic schists. Occasionally there are lenses of highly silicified slate, such as has been encountered in the Howie (Union County), Thompson (Stanly County), and Iola (Montgomery County) mines.

The general strike of these formations is in a northeast-southwest direction, with a dip to the northwest and southeast.

The ore deposits in the slate belt consist of both veins and lodes.

There is a greater concentration of vein deposits along the western side of the slate belt. The most important lodes or mineralized zones occur along the eastern border, especially in Montgomery County. However, some quartz veins are found in Montgomery County as at the Iola Mine. Few of the mines in the slate belt have been worked to depths greater than 500 feet.

The chief gangue minerals occurring in the mines of the slate belt are quartz, pyrite, chalcopyrite, siderite, calcite, and in some instances bornite and rhodocrosite. In the upper zones of the veins and lodes, where oxidation has taken place, the sulphides have been altered to oxides, and the gold is in the free state.

Often the slaty walls of the veins are impregnated with auriferous sulphides in some of the lode deposits; however, there are apparently no distinct quartz leads, but the slates themselves contain small crystals of pyrite and often these zones have considerable width, as at the Russell, Howie, Keystone, and Bonnie Bell mines. At the Thompson Mine in Stanly County there are layers of pyrite conforming more or less to the schistosity in addition to small crystals of pyrite included in the slate. These latter crystals are distributed irregularly through the slate. The pyrites, on decomposing, free the gold.

The source of the ores of this slate belt is apparently from deep-seated granitic magmas, but occasionally there seems to be some genetic relationship between the basic intrusions and the ores.

The rocks of the slate belt are pre-Cambrian in age; the exact age, however, is not definitely known due to the lack of fossils. When geologic studies were first made of this slate belt, it was thought that fossils had been found. To these, Emmons gave the name *paleotrochis major* and *paleotrochis minor*. Later, however, these so-called fossils were found to be spherulites. A slate outcrop north of Mt. Gilead in Montgomery County includes numerous masses of material which have a decidedly peaty appearance.

The Triassic deposits to the east of the slate belt, especially the basal conglomerates, contain placer gold. Therefore, the age of the deposits in the slate belt must be early Paleozoic.

ORE DEPOSITS AND MINES IN THE SLATE BELT

In the extreme northeastern section of the slate belt, near the Virginia line, there is a series of copper deposits which contain low gold values. However, the principal work carried on in this section was essentially for copper. This mineralized belt is approximately 10 miles long and from 2 to 3 miles in width.

The veins conform more or less to the schistosity of the slates, and vary from a few inches to 14 feet in thickness. The veins consist largely of quartz carrying chalcocite (gray copper) and bornite, with some red copper oxide, carbonates, and rarely native copper.

The principal mines in this area are the Royster (or Blue Wing), Holloway, Mastodon, Buckeye, Pool, Gillis, Copper World, and Yancey. None of these mines have been worked in recent years. All of these mines are found in Granville and Person counties.

In Alamance, Orange, and Chatham counties gold has been found at a number of localities, but no section has been worked to any great extent. The most important development in the county was the Robeson Mine, 12 miles northwest of Chapel Hill. The ore-body is a quartz vein, varying from 6 inches to 22 inches in width. The strike is northeast and the dip 30° to the northwest. In 1895 a prospect shaft was sunk to a depth of 30 feet, at the bottom of which the vein was 22 inches in width. The quartz vein is cellular, vitreous, to saccharoidal, and has a good appearance. Assays showed values from six to fifty-two dollars per ton.

MOORE COUNTY

The mines in Moore County are confined to the northwestern part of the county, on the northwest side of the Triassic basin.

The *Bell Mine* is located in the northern part of the county, 8 miles northwest of Carthage. The country rock is a chloritic schist, having the appearance of a metamorphosed eruptive. The country rock is cut with small calcite seams. The strike is N. 55° E. and the dip 75° N. W. The upper part of the schist, as shown in the shaft, bends over with the slope of the hill due to creep.

The paystreak is reported to be from 4 to 8 inches wide lying against the foot wall, and that $1\frac{1}{2}$ to 2 feet of the material on the foot wall side was mined and milled, yielding as much as $30 per ton. The entire vein might be said to average 4 feet in width, and shows $12 per ton. A rigid sampling of the entire ore body on the 75-foot level was made, the assays showing from $11 to $14.50 per ton.

In the upper zone there were sulphides present, and the free gold was very "leafy," which caused great difficulty in working the ores by the ordinary methods of amalgamation.

The *Burns Mine* is also known as the Alred, and is situated 11 miles northwest of Carthage on Cabin Creek.

The country rock is a sericitic, chloritic schist, at times silicified. The strike is N. 20° E. and the dip 55° N. W.

The schist is filled with quartz stringers, and it is difficult to say what is ore, as the rock is everywhere auriferous but not always capable of being worked profitably. The ore was mined in large open cuts, 20 to 100 feet in width, to a depth of about 50 feet. These open cuts extend along the strike for a distance of about one-fourth mile.

The cuts are scattered about promiscuously, without much evidence of connection or relation, and are usually very irregular in outline.

Iron sulphides occur in the schist but none of these were treated. Quartz veins intersect the schist in all directions but are apparently barren. The average of the ore is $2.50 to $3 per ton, with occasional intervals of high-grade ore. Due to the nature it is impossible to tell the high-grade material except by panning or assaying.

The *Cagle Mine* is situated about three-fourths mile north of the Burns; the country rock and ore are similar. The strike is N. 27° E. and the dip 55° N. W. A series of shafts has been sunk on the dip to a depth of 160 feet. The ore ran from $4 to $48 per ton, with an average of about $25.

The *Clegg Mine* is one-fourth mile west of the Cagle on the west side of Cabin Creek. The character of the ore is similar to that of the Cagle, the ore body larger but of relatively lower grade. This mine has also been worked by open cuts.

The *Brown Mine* is on the northeast edge of this district, and has been worked to a depth of 40 to 50 feet over a distance of 300 yards. The dip is very flat, and the ore-body is about 3 feet thick. The paystreak is a comparatively narrow seam of rich quartz.

Other mines in the county, which have shown production but are of little importance, are the Grampusville, the Bat Roost, and the Shields.

RANDOLPH COUNTY

The *Hoover Hill Mine* is situated about 17 miles east of south from High Point, on the Uharie River. This mine has been worked spasmodically for a number of years.

The country rock is a decomposed basic eruptive, which is brecciated and includes fragments of hornstone. Masses of this hornstone contain sulphides which resemble the ore of the Silver Valley Mine. More or less definite belts of this rock, apparently a quartz porphyry, are intersected by a series of small reticulated quartz veins varying in width from an inch to a foot or more; this material constitutes the ore. The weathered portion was extraordinarily rich, and gave the mine its early fame. The general strike of the belt is northeast and the dip 30° to 60° southeast. The ore bodies are intersected by pyroxenic dikes.

The principal ore body is known as the "Briols" shoot. This shoot was worked to a depth of 350 feet. It is reported that the ore body at this depth was large, its width 12 feet and its length 70 feet or more, with values $8 to $10 per ton. At a short distance from the Briols shoot there are six other ore bodies lying quite close together, which have been worked from one shaft.

The *Jones* or *Keystone Mine* is located 18 miles east of south from Lexington. The country rock at this mine is quite similar to the brecciated porphyry found at the Hoover Hill Mine. However, it is much more schistose. Near the surface the rock is very soft and decomposed, the iron sulphides in it being oxidized into limonite. Due to this oxidation the whole mass of rock is stained a reddish-brown. The strike of the formation is N. 45° E., dipping about 80° N. W. In addition to the sulphides, the highly schistose rock is filled with fine quartz stringers. The ore-bodies consist of the quartz-pyrite mass.

Gold is more or less distributed throughout the decomposed portion but is confined to certain well-known belts which are more richly charged with it. There are zones or horses charged with finely disseminated pyrite, which are slightly altered and are more or less firm. These firm zones have been avoided due to the difficulty of mining and milling. Two of these belts have gained especial prominence, one 50 feet and the other 110 feet wide. There are numerous openings over the entire tract, and all of these openings show value. The barren portions are easily recognized and can be avoided during mining operations. The mine may be considered as a series of open quarries. It is stated that the character, the occurrence, and the distribution of the ore are such that new bodies equal in value to those that have been worked may be exposed at any time. The soft disintegrated condition of the rock permits mining at a very cheap rate.

The ore, according to assays, shows values from $2 to $28 per ton with an average of about $3.

The upper portion of the deposit was worked for a number of years profitably, but as depth was reached and sulphides were encountered the values apparently decreased; therefore it was soon abandoned. In the early days a combination of hydraulicking and milling process was applicable. The nearest water is the Uharie River some two miles distant.

The *Uharie Mine* is only a short distance to the northeast of the Russell Mine, Montgomery County, just over the line in Randolph County. The formation at this mine is quite similar to that at the Russell Mine. The work at the Uharie Mine, however, has been underground, the deeper shaft being 170 feet in depth. It is reported that the rock carries 1½ per cent pyrite.

The other mines of less importance in Randolph County are the Parish, Lafflin, Delft, Winningham, Slack, Davis Mountain, Sawyer, and Winslow. None of these mines have been developed to any great extent.

The *Alred Mine* is 10 miles northeast of Asheboro. This mine consists of a number of open cuts distributed for $\frac{1}{4}$ mile along a belt 400 feet wide, or more. The country rock trends in a northeast direction, and is apparently an altered slate or schist. In 1933, considerable work was done at this property in prospecting and testing to prove the value and extent of the ore deposit. A ten-stamp mill stands on the property, but has operated intermittently during the past few years. Considerable trouble was encountered in trying to save the gold due to its fineness and to the nature of the material, more or less clayey, in which the gold occurs. To overcome this problem, various methods were tried out, including an electrical method, but apparently without success, as they were soon abandoned. The best information obtained as to the value of the ore shows a gold content of less than $1 per ton. However, the limonite pseudomorphs ran as high as $12 to $15 per ton.

DAVIDSON COUNTY

The principal mines in Davidson County are the Emmons or Davidson, Cid, Silver Hill, Silver Valley, Conrad Hill, and Welborn. All of these mines are fully described in Bulletin No. 22, "Cid Mining District of Davidson County," by Joseph E. Pogue, Jr., 1910. This publication can be had from the North Carolina Development of Conservation and Development.

MONTGOMERY COUNTY

There are several mines in Montgomery County, some of which are deep mines, which have had a very prominent history in the mining industry of the State. Probably the most important group is located in the extreme northwestern corner of the county in the vicinity of Eldorado.

The *Russell Mine* is situated three miles north of Eldorado near the Randolph County line. This mine has been worked more extensively than any other mine in this vicinity; therefore it permits better opportunity for study than the other mines.

The country rock at the Russell Mine is an argillaceous slate, both the soft and decomposed type with silicified phases; it also contains small quartz stringers. The silicified slate showed coatings of calcite, while the softer material contained small irregular calcite veins. Apparently the slate is somewhat calcareous. The dip and strike are variable. The bedding and cleavage plains usually coincide, but in cases particularly where the bedding is at a small angle to the horizontal, they do not coincide, the cleavage being much steeper. The cleavage plains are also marked by groovings. Pyrite to the extent of 2 to 4 per cent is present in the slates. The pyrite crystals occasionally are one-eighth inch in diameter and are disseminated in irregular stringers, lenticles, and coatings. The entire belt is gold-bearing, but only certain zones are rich enough to warrant mining operations.

To one unfamiliar with the North Carolina slates it is hard to distinguish the difference between the rich and poor portions of the gold-bearing zones. Even those familiar with the ores are controlled by constant panning.

All of the mining so far done at this property was principally by open pit with some drifting. The ore over the entire width or "leads" showed values of approximately $2 per ton. However, there were rich zones 4 or 5 feet in width that carried values up to $16 per ton. An occasional streak was found so rich that the miners carried the ore out in kegs. These larger streaks showed values up to $332 per ton.

Six of the belts, within a distance of 2,000 feet across the strike, have been worked, the largest or "Big Cut" was worked for a distance of 300 feet in length, 150 feet in width, and 60 feet in depth.

In 1894, a 40-stamp mill was at the mine, and preparations were made to erect a cyanide plant for the treatment of the ores. No information is available as to the success or failure of this plant.

The *Coggins (Appalachian)* gold mine is located in the northeastern part of Montgomery County, 1½ miles north of Eldorado.

The country rock of this area is composed of argillaceous slates or schists, probably derived from land waste, which have varying amounts of tuffaceous material. Cutting these rocks are diabase dikes up to 6 feet in width. The strike is N. 40° E. and the dip 75° to 80° N. W. The slates are both soft and silicified, and carry quartz stringers which vary in width from very narrow up to 10 feet or more The mineralized zones vary in width up to as much as 50 or 60 feet. Free gold occurs in the quartz seams, but at depth the ore is principally a sulphide. Due to this, the free gold can not be saved by amalgamation.

There is a great variation in the values in the ore deposits, and there seem to be well-defined ore shoots, which have a lenticular structure, that are richer than the average of the vein. In some cases the walls of the vein are well-defined, but assaying is required to determine the extent the vein could be worked profitably. It is almost impossible to distinguish between the rich and poor sections of the vein.

Numerous assays of the ore show values from $1 to $6.77 per ton; the average is not definitely known. One run of ore from the 200 to the 250-foot level gave values of approximately $20 per ton. The ore delivered to the mill from this level showed assays up to $53 per ton.

The mine has been worked to a depth of 250 feet, and drifts have been sent out at the 50-, 100-, 200-, and 250-foot levels.

The mine was partially unwatered in 1933 and '34, but the values apparently did not justify further exploration, as it was soon abandoned. However, since that time workmen at the property stated that the value of the ore was approximately $10 to $15 per ton and apparently occurred in large quantity.

The 40-stamp mill which stood on the property for many years has been practically torn down and the material sold to near-by mine operators.

The *Riggon Hill Mine* is situated 3 miles east of Eldorado. The ore-body consists of a quartz vein, reported to be 2½ feet in thickness, lying in part conformable with the schistosity of the country rock. The vein has been opened by a prospecting shaft 100 feet in depth. Rich ores, both in gold and silver, are reported from this vein.

The *Steel Mine,* and its extension, the *Saunders,* are two miles southeast of Eldorado on the east side of the Uharie River. The country rock and the ore of these mines are quite similar to those of the Russell Mine.

The ore-body varies from 9 to 12 feet in thickness and occasionally reaches 20 feet. The strike is N. 25° E. and the dip 70° N. W.

The most valuable part of the deposit consists of stringer veins which cut the rock, generally conforming with the schistosity. These stringers, approximately parallel, have a combined thickness of 15 inches or more, occasionally reaching 3 feet. There is considerable free gold, but a large part of it is associated with the sulphides; therefore a smelting treatment is necessary.

The ore has been mined to a depth of 220 feet, and has

shown values up to $9,800 per ton. These values justify the early reputation of the mine for richness. The mine was last operated in connection with a 40-stamp mill. At depth the sulphides were encountered and had to be concentrated for treatment.

The *Moratock Mine* is situated 8 miles south of Eldorado. The country rock is a highly silicified quartz porphyry and brecciated tuff. The porphyry contains pyrite crystals distributed irregularly along small seams and in minute cavities or geodes. The sulphides carry some copper. The main formation is intersected by several small quartz fissure veins less than one inch in thickness, which are reported to be very rich in gold.

This mine was operated first in 1892 by open pit methods. The principal opening was confined to the pyritic quartz-porphyry mass. The assays on the ore ran less than $1 per ton; therefore it was soon abandoned.

The *Sam Christian Mine* is situated 12 miles east of Albemarle. This mine is known especially for the remarkably large and fine nuggets found. Forty were found which gave a total weight of 4,200 pennyweights.

The mine was last operated in 1893. Water had to be pumped from the Yadkin River about 2½ miles distant.

The *Beaver Dam Mine* is located 2 miles northeast of the junction of Beaver Dam Creek and the Yadkin River. The gold found at this mine occurs in a gravel 2 to 4 feet thick, overlain by alluvial deposits 5 to 15 feet thick. The bed rock is a decomposed silicified schist. A greenstone, large and extensive, occurring on the property, contains a small amount of pyrite which shows assays up to $2, or better, per ton.

The *Black Ankle Mine* is situated 14 miles northeast of Troy. The gold occurs in a mineralized zone in a highly schistose rock belonging to the slate group. This deposit

Fig. 8. Prospect Shaft, Furr Mine, Cabarrus County.

Fig. 9. Gold Recovery Plant, Black Ankle Mine, Montgomery County.

was discovered in 1928.   Several attempts have been made to operate this property but so far have been unsuccessful.

The workings consist of a pit 225 feet in length, 120 feet in width, and 50 feet in depth.   In addition, on the west bank of the pit there is a shaft 112 feet deep.   It is reported that a drift 150 feet in length crosscuts the ore body at depth.   The gold usually occurs associated with the quartz stringers more or less paralleling the schistosity of the rock.

The gold is extremely fine, and all the methods so far attempted to recover it have been unsuccessful.   The material has been treated by washing, amalgamation, cyanidization, and by the Lewis centrifugal processes.   It has been claimed that most of the gold was lost on account of its fine subdivision and the nature of the slime produced by the semi-kaolinized volcanic ash when washed.   Engineers have reported on the property, but the general average of the assay values show less than 50 cents per ton.   It is likely that only the extremely rich streak, very narrow, could be worked profitably.

The *Iola Mine* is situated 3 miles to the west of Candor. This mine occurs in a section of the country little prospected previous to 1900.   When prospecting began, the Iola vein was discovered.

The ore deposit consists of narrow veins, siliceous pyrite zones, and aggregation of stringers and lenses with unaltered schists between.   The country rock is a thinly laminated schist for the most part, but graduates into a more massive rock, apparently an andesite.   The veins are crossed by dikes of diabase and other basic rock and are faulted. The vein is from 1 to 8 feet in width and has been developed along the strike for a distance of 2,000 feet and to a depth of 260 feet on the dip.

The *Montgomery Mine* is located 2½ miles west of Candor.   The country rock is a hard siliceous greenstone slate which has been much broken by jointing. It is cut by two basic dikes, apparently diabase, one of which has faulted the vein while the other has not.

There are two veins on the property varying from 1 to 6 feet in width with an average of about 2 feet. The underground development work reaches a depth of 390 feet with drifts 800 feet in length along the strike. The veins were opened at the 100-, 150-, 225-, and 300-foot levels; the shaft continued to 90 feet below the 300-foot level.

The value of the ore is not definitely known, but, according to the amount of work accomplished, must have averaged sufficiently high to have been a profitable operation. The extraction was 94 to 97 per cent of the values.

The *Carter Mine* is located about 3 miles west of Star. It is one of the older mines in the State. The country rock is a basic schist, an altered andesite, and the ore is quite similar to that of the Iola Mine. The ore zone is from 1 to 2 feet in width and consists of quartz stringers interlaminated with the country schist. Drifts along the vein showed that the ore pinched and swelled considerably. A shaft sunk, in 1913, 50 feet in depth encountered several faults. The early mining consisted of a shaft 65 feet in depth, which encountered ore said to average from $6 to $7 per ton. At one place in the south end of the drift, ore averaging from $16 to $80 per ton was encountered.

Other mines in Montgomery County which have shown gold production are the gravel mines found on the west flank of the Uharie Mountains, between them and the river.

The localities best known and most worked are the Bright, Ophir, Dry Hollow, Island Creek, Deep Flat, Spanish Oak Gap, Pear Tree Hill, Tom's Creek, Harbin's, Bunnell Mountain, Dutchman's Creek, and the Worth.

Considerable trouble was encountered at all of these properties in recovering the gold, due to the tenacious clay overlying the gravel. Also the scarcity of water hampered operations. There is apparently considerable gravel at the heads of these shallow valleys which has never been worked.

STANLY COUNTY

The *Haithcock* and *Hern Mines* are located 2½ miles northwest of Albemarle. The country rock is a schist, apparently an altered slate. The ore occurs as quartz veins 2 to 6 feet in thickness, more or less conformable with the schist. In 1933 and '34 considerable prospecting was done along the outcrop of these veins, and a 5-stamp mill was erected. The gold occurred as free gold in the quartz and was readily amenable to amalgamation. At several places along the vein the gold occurred in sufficient quantity to be easily discernible without the aid of lenses. The deepest shaft did not extend below water level. These properties have possibilities, provided an equitable contract could be had with the owners.

The *Parker Mine* is located at New London about 7 miles northwest of Albemarle.

The country rock is the typical Monroe type intruded by successive flows of greenstone porphyry and more basic eruptives, in part brecciated. In places the greenstone is sheared into nearly vertical schistose masses. The shafts have revealed two flows from 2 to 3 feet in thickness lying almost horizontally.

Numerous quartz stringers, varying from very thin to 18 inches or more in thickness, intersect the country rock in all directions and at all angles. In addition to these small stringers, there are larger and more persistent veins. A peculiar property of these quartz veins is that they are usually imperfectly crystallized and often cellular.

During the process of weathering, the rock has been decomposed to depths of 10 to 20 feet, the breaking down of the quartz veins having caused an irregular distribution of the gold through this decomposed rock. In the low places there has been a greater concentration of the quartz fragments and gold. However, a great many of these deposits have been completely hydraulicked and the gold recovered.

The principal yield of the gold at this mine has been from the old gravel channels. Most of the gold was and is very coarse, and a great number of nuggets have been found, the largest of which weighed 8 pounds, 3 ozs., and 2 dwts. In 1935 other nuggets have been encountered, which weighed from 1 to $2\frac{1}{2}$ pounds each.

The value of the gravel, as shown by numerous pannings and assays, ran from 40c to $3.20 per yard. Even though over $200,000 in gold has been produced from the gravels on this property, they are by no means exhausted. Pits which have been sunk on the placer material have revealed the greenstone flows underlain with other gravel which has not been hydraulicked and may warrant further development. However, this could only be determined by further prospecting and sampling.

Three shafts have been sunk on the property, the "Ross shaft," 120 feet in depth; the "Crib shaft," 80 feet in depth; and another shaft said to be over 100 feet in depth. These old shafts have not been unwatered for several years, and little is known as to the value of ores encountered in them. However, miners who have worked in these shafts state that some very high-grade ores were encountered and have been by no means exhausted.

In the early part of 1933 the property was optioned by E. C. Gallagher of New York. The State Geologist supervised the sampling of the deposits, and approximately 7 tons of material were taken. About 400 pounds of the quartz vein, 600 pounds of the schist vein, and $6\frac{1}{2}$ tons of the placer material were concentrated to 1,500 pounds and shipped to the Columbia School of Mines for testing. Instead of running the samples separately as taken from the property, the ore was milled and gave an average of better than $3 per ton. However, since the material was mixed before milling, the source of the gold was not known, whether from the quartz vein, schist zone, or placer. Advice was given to re-sample the deposit and run the material separately, but this was not done. A small mill was erected to handle the surface material but proved unsuccessful and was soon abandoned. It was stated by the owners that the mill did not recover the values. This was demonstrated by the fact that a G-B machine recovered values up to $9 per ton from the tailings, a great deal of which was nugget gold.

In 1935 the North Carolina Mining Corporation, Washington, D. C., assumed control of the property. A tunnel was sent into the hillside for a distance of 250 feet to intersect any veins that might be found cutting the country rock. At a distance of 150 feet from the adit a quartz vein varying from 6 to 18 inches in width was encountered. In addition several small stringer veins were also found. A shaft about 15 feet in depth was sunk on the quartz vein encountered in the tunnel. From this small shaft about 15 or 20 pounds of gold were taken, most of which was in nuggets up to 2½ pounds in weight.

The property has been sold to another group which plans to develop it on a large scale.

The *Crowell Mine* is located one mile east of New London. The country rock is a silicified, sericitic, and chloritic schist, containing finely disseminated pyrite. The ore body differs little from the country rock, as both are auriferous. The ore body, however, is said to be from 4 to 7 feet in thickness and will pay to work as a whole at times. The "paystreak" is much narrower and often becomes very thin. A shaft on the property was sunk to a depth of 125 feet, and a crosscut was sent eastward, which encountered the vein. Little drift work was done on the vein. On the west side of the shaft 2 or 3 pits have been sunk on the west vein to a depth of 10 or 15 feet. Samples from these pits showed values from $2 to $6 per ton over a width of 4 to 6 feet. When the shaft was sunk, it was the intention of the operators to crosscut to the west vein, but this was never accomplished.

The property has been purchased by Mr. Cassidy, of Charleston, W. Va., and he is crosscutting the mineralized zone to a depth of 30 feet. As is readily shown by panning, gold is revealed almost the entire length of the crosscuts. The present operators plan to erect a gold recovery plant at an early date.

On the north end of the Crowell mining property considerable exploration work was done to determine the extent and the value of the quartz stringers. These quartz stringers vary from very thin to 3 or 4 inches in width and often show gold in considerable quantity. However, to date none of the veins have shown gold in sufficient quantity to make a profitable operation. Some of the local miners worked these veins spasmodically as tributors. On the surface, near and along the veins, very beautiful nuggets have been occasionally encountered.

The *Barringer Mine* is located 4 miles southeast of Gold Hill. The ore from this mine was reported to be very rich; values up to $500 per ton or better are reported. It is very deceptive, however, and the value can only be determined by careful panning or assaying. No work has been done on this mine during the past 40 years.

The *Crawford (Ingram) Mine* is situated 4 miles east of Albemarle. This placer deposit was discovered in August, 1892.

The auriferous gravel is situated in the valley of a small stream and is overlain with 2 to 4 feet of overburden. The thickness of the gravel varies from $1\frac{1}{2}$ to 2 feet, or better, in the center of the basin, thinning out towards the edges. The width of the gold-bearing gravel is about 250 feet.

The bed rock in this area is a slate of the Monroe type lying slightly synclinal. The gravel is composed of the angular fragments of the slate and the white quartz from the quartz veins cutting the slate, which are bound together more or less with the clay matrix. The stream in the center of the little valley is not sufficient for washing the material.

Two particularly large nuggets were found, one weighing 8 pounds and 5 ounces, on April 8, 1895, and another weighing 10 pounds, on August 22, 1895. A great deal of the gravel was worked directly by rockers, sluice boxes, and other simple methods, by tributors, who paid a royalty to the owners.

The source of the gold in this placer material is not definitely known, as the quartz veins on either side of the valley have been explored considerably, but no coarse gold has been found. The quartz fragments on the hillsides have been crushed and panned but have shown low auriferous values.

CABARRUS AND ROWAN COUNTIES

The mines in the Gold Hill group are fully described in Bulletin No. 21, "The Gold Hill Mining District of North Carolina," published in 1910. A copy of this publication can be had from the North Carolina Department of Conservation and Development, Raleigh, North Carolina. In addition to the Gold Hill, other mines described in this publication are Union Copper Mine, the Whitney mines, the McMakin, Isenhour, and Barringer mines.

The *Mauney Mine* is located 1½ miles southwest of Gold Hill, on the opposite side of Little Buffalo Creek. The country rock is a schist which strikes N. 25° E. and dips 75° N. W. The schists are very soft and contain but very little quartz. The gold-bearing portion is considered to be 4 feet in width, which showed values from $2 to $6 per ton. In

1894 a prospecting shaft was sunk to a depth of 70 feet, or just below water level. The upper soft brown ores were worked to some extent.

The *Rocky River Mine* is located 10 miles southeast from Concord, on the waters of Rocky River. The country rock is a sericitic and chloritic schist, very soft at times while at others very highly silicified, with all grades in between. The strike of the formations is N. 20° E., and the dip of the schistosity 70° N. W. There are several parallel quartz veins which lie more or less paralleling the schistosity but often cut it at slight angles.

The chief explorations carried on at this mine were done in 1895, and consist of 2 shafts, 130 feet and 200 feet, respectively, in depth. At the 80-foot level the quartz was from 14 inches to 3 feet in thickness; in the southwest drift, about 50 feet from the shaft, it pinched to a few inches. At the 130-foot level the vein was lost entirely.

The vein matter is principally quartz with some carbonates, apparently calcite. The quartz carries sulphides, chiefly pyrite with smaller amounts of galena, sphalerite, and chalcopyrite. According to tests, about 50 per cent of the gold milled was free milling. Apparently there are three grades of ore: (1) rich in galena, which carries much of the gold, a smelting ore; (2) rich in pyrites, an ore to be treated by amalgamation and flotation; (3) lean ore.

Assays on the ore show values from $6 to $68 per ton with an average of about $15 per ton. However, some of the sulphides show values up to $30 per ton. The quartz veins, with the schist in between, and to some extent the schists on either side of the veins are mineralized and if taken together would possibly show a large body of low-grade ore. If the ore were worked in this manner, it is possible to work it profitably.

The *Allen Furr Mine* is 11 miles southeast of Concord

Fig. 11. Showing Six-Foot Fissure Vein at Allen Furr Mine. Notice Small Fault Near Top of Picture.

Fig. 10. Surface Equipment at Allen Furr Mine, Midas Mining Co., Winston-Salem.

and 23 miles east of Charlotte. The mineral property consists of 20 acres, and is about 1,500 feet in length along the vein.

The country rock is a slate of the Monroe type, which strikes west of north. The vein is formed along a fracture, and consists of quartz with varying percentages of sulphides up to 25 per cent. The sulphides are principally pyrite, galena, and sphalerite. The values in the vein show 2 to 18 per cent lead, 3 to 11 per cent zinc, less than .25 per cent copper, 1.2 to 1.49 ounces of silver, and gold from $5 to $32 per ton with an average of approximately $12.

A shaft has been sunk on the vein to a depth of 90 feet. At the 45-foot level a drift 225 feet in length was sent out on the vein which showed a width of 30 inches. At the 90-foot level a second drift was begun which showed a vein between 5 and 6 feet in thickness that carried values as follows: 5 per cent lead, 2 per cent zinc, 6 ounces of silver, and $14 in gold. In the summer of 1935 the mine was operated by Mr. Terry of Charlotte, the ore hauled to the flotation mill on Rocky River and the concentrates shipped to the American Metals Company, Carteret, New Jersey. Mr. Terry stated that the concentrates averaged about $40 per ton.

The *Snyder Mine* is located 8 miles southwest of Mt. Pleasant.

The country rock consists of a slate, of the Monroe type, considerably sheared. The shear zone may be classed as a schist.

In the summer of 1935, under the direction of E. L. Hertzog, Spartanburg, South Carolina, a shaft was sunk to a depth of 137 feet on the vein. No drifting was done. Near the surface the vein width, including mineralized schist partings, was about 4 feet. The chief gangue mineral is white and rose rhodochrosite. At a depth of 137 feet the vein

Fig. 12. Sluice Boxes, Reed Mine, Cabarrus County.

Fig. 13. Remains of First Chilean Mill Erected in N. C. at Reed Mine.

had pinched to 6 to 14 inches in width. The average gold content in the upper zone is from .35 to .4 ounces per ton. However, the values steadily decreased on the average as depth was reached.

Mr. A. L. Nash, Salisbury, North Carolina, submitted the following information on this property:

When the shaft was started, the vein was 3 feet in width and the assay values were from $63 to $66 per ton. Samples from the dump showed values $2.45, $9.20, $24, $24.15, and $42. At the 90-foot depth a sample taken 5 feet across the vein showed $18.55. Samples assayed at Gold Hill, wall rock 35c, or 3-foot section $45.50, quartz with sulphides $12.60, quartz $7.70. Samples assayed at Charlotte, quartz $54, quartz and schist $24.63.

Mr. Nash stated that he believed that the ore would average between $10 and $20 per ton from the vein, which averaged from 3 to 4 feet in width.

The *Harkey Mine* is located 5 miles southwest of Mt. Pleasant. The country rock is a diorite. The vein is from 6 to 24 inches in width, and is composed of quartz, chalcopyrite, pyrite, and marcasite. The shaft is 61 feet in depth, and the drift at the bottom of the shaft showed a vein from 24 to 30 inches in width. The property, in the summer of 1935, was being developed by Mr. A. L. Nash, Salisbury, North Carolina. Mr. Nash stated that the assay values were from $4 to $67 per ton with an average of $34 per ton.

The *Reed Mine* is located 1½ miles southeast of Rocky River near Georgeville. It is the site of the first discovery of gold in North Carolina. Over 150 pounds in gold have been taken from the property in nuggets, some of which weighed 15 to 28 pounds each. This mine was one of the first gold mines operated in the State, and it is estimated that from 1804 to 1846 over a million dollars in gold were taken from the property. The last nugget of any consequence was discovered on April 11, 1896, which weighed 10.072 pounds. This discovery caused a revival of interest, but it was short-lived.

The country rock at this mine is a chloritic schist accompanied by a large body of greenstone. The country rock is intersected by numerous quartz veins varying in thickness from 4 inches to 3 feet. The gold-bearing veins were prospected to a depth of 120 feet. The values were disappointing, and the property was soon abandoned.

In the summer of 1934 some placer work was carried on along the stream and the hillside, and several small nuggets were found. The old tunnel cutting the vein at depth was opened up and some sampling done. The results were apparently not satisfactory as it was soon abandoned.

The *Phoenix Mine* is situated 7 miles southeast of Concord. The country rock is a schist accompanied by a large mass of diabase. The quartz veins are confined principally to the diabase. The main Phoenix vein strikes N. 70° E., and dips 80° N. W. It varies from 12 inches to 3 feet in thickness. This vein was extensively and successfully worked for several years under the direction of Capt. A. Thies, but operations ceased in 1889.

The ore shoot, which is 300 feet long and pinches to the northeast, has been worked out from the 100-foot to the 425-foot level. The shaft was sunk to 485 feet on the dip of the vein, but no drift work was done at this level. The vein in the shaft averages 30 inches, but the rich paystreak lying on the hanging wall is only from 2 to 3 inches thick. It is believed by some that, if the vein were drifted on at the 425-foot level, the 300-foot ore shoot would be encountered again. Another ore shoot, the Big Sulphur, is situated 300 feet southwest of the other shoot and has been worked to a depth of 180 feet. The ore in the bottom of the shaft on this shoot is said to be 14 inches thick. The work accomplished by Capt. Thies was confined to the 300-foot shoot. The ore was quartz carrying 3 to 6 per cent sulphides, such as pyrite, chalcopyrite, and traces of galena. In addition to the quartz gangue, there is also some barite and calcite. The assays show only ½ to 3 per cent copper. The mill yield was $10 per ton in gold, and the concentrates ran $30 per ton. The chlorination process was introduced at this mine in 1880; at first it was the Mears process but later developed into the Thies process. In 1889 the mill and chlorination plants were dismantled.

The *Buffalo Mine* is situated one mile northeast of Rocky River. The country rock is a slate which strikes N. 55° E., and dips 80° N. W. The main vein is a quartz, 5 feet in maximum width. The slates have been cut in a shaft and crosscut over a width of 25 feet. The slates contain pyrite, and show $3.50 per ton in gold.

The *Nugget* or *(Biggers) Mine* is situated 12 miles southeast of Concord. The country rock is an argillaceous slate, which is intersected by basic dikes. In 1894 to '96 the gravel channels were worked to some extent by hydraulicking. The gold is usually quite coarse and in nuggets. The quartz veins, carrying galena, were explored superficially but never developed.

The *Crayton Mine* is located 2½ miles north of Georgeville. The country rock is principally a slate which strikes N. 35° E. The veins are quartz with pyrite, which vary in width from 2 to 6 feet. The gold occurs as minute grains in the quartz. The lode is displaced by overthrusts. A mill test reported $6.47 per ton.

In 1933 this property was explored with a shaft 40 feet in depth. The vein is reported to be 6 feet in width at the bottom of the shaft. A group of doctors from Gastonia erected a small 10-stamp mill at the property, but the ore was not amenable to straight amalgamation. The plant was soon abandoned.

The *Tucker* or *California Mine* is one mile south of the Phoenix. It was last worked in 1884 by a shaft 175 feet in depth with levels 117 feet in total length. The quartz vein was 15 inches in width and showed values of $15 per ton. In 1882 the Platnner chlorination process was introduced at the property, but was later superseded by the Mears process.

The *Quaker City Mine,* which is 3 miles north of the Tucker, was prospected with 3 shafts, the deepest of which was 80 feet. The vein is reported to be from 3 to 5 feet in width.

The *Barrier, Furness,* and *Gibb mines* adjoin the Phoenix Mine. The Faggart is 3 miles to the northeast, and the Barnhardt is 1½ miles east of the Faggart.

The *Pioneer Mills* group of mines is situated 13 miles south of Concord. The granite in this locality is accompanied by large masses of basic eruptive rocks. This last group of mines has not been worked since the Civil War, but some exploration work was carried on in 1930-31. Apparently nothing of value was uncovered, as the properties were soon abandoned.

UNION COUNTY

The gold mines in Union County are principally confined to the western side, west of Monroe. The ore is principally sulphides, both auriferous and argentiferous. Copper is present but never in quantity.

The *Crowell Mine* is situated 14 miles north of Monroe in the extreme northwestern corner of the county. The country rock is a schist, which strikes in a northeast-southwest direction and dips from 40 to 45° northwest.

There are three veins which vary in thickness from 1 to 4 feet. The deeper shaft is 80 feet with drifts 50 feet northeast and 60 feet southwest. The values vary from $6 to $86 per ton with an average of about $20.   At times the silver content is high.

The *Long Mine* is ¾ mile southwest of the Crowell and is apparently in line with the Crowell veins.   The strike is N. 50° E., and the dip 85° N. W.   The vein is similar to that at the Crowell Mine, and the vein matter is quartz and schist which carry the sulphides, especially pyrite, with some galena, sphalerite, and chalcopyrite.   Calcite and siderite occur as gangue minerals.   The vein is reported to be from 30 inches to 4 feet in width.   The average gold content is not known.

The *Moore Mine* is 3 miles southeast of the Long.   The country rock is a chloritic schist, which strikes N. 50° E. and dips steeply to the northwest.

The mine was opened by a shaft 80 feet in depth.   The vein is reported to be 5 feet in thickness, approximately conformable to the schistosity of the country rock.   A four-inch paystreak, composed mainly of calcite, carries free gold and follows the angle of the wall.

The quartz carries pyrite, chalcopyrite, galena, and sphalerite.   The sulphides also extend into the wall rock. The schists also are intersected by narrow seams of calcite.

Assays on the ore show values of $9.79, $23, and $240.88 per ton.

The *Stewart Mine* is located 1½ miles southwest of the Moore, on Goose Creek.   The country rock is an argillaceous and sericitic schist with a strike N. 55° E. and a dip 85° N. W.   The schists vary from very soft to highly silicified types.   The ore bodies consist of certain amounts of the country rock, impregnated with auriferous quartz stringers, sulphides, pyrite, and galena.   There are said to be three parallel ore belts, the Asbury, 4 to 18 inches in width; the Miller, 5 feet in width; the Jake, the width not known.   The veins have been explored to a 285-foot depth.

The seams of ore are very narrow, but many of them (and it is a question whether the whole series is not one and the same ore belt) have several production channels. The widest vein is 12 feet, but is not externally promising. The assays of the ores showed $9.20, $17.19, $78.70, $47.93, $50.19, and $240.78. All of the assays showed silver and lead. A very rich shoot of ore was found but was lost; a great deal of costly work was expended in searching for it, but it was never found again.

The *Lemmonds* or *Marion Mine* is apparently a southern extension of the Stewart. The vein is irregular in size, sometimes widening out from a few inches to 6 feet. The vein consists of quartz richly charged with brown zinc and galena. An average of the ore showed a value of $9.90 per ton. The pure sulphides showed values as high as $730.00 per ton.

The *New South Mine* is 1½ miles northeast of the Stewart. The country rock is a slate laminated and thoroughly altered. At depth, however, the alterations and the blue slates came in. The work done to the 25-foot level was satisfactory and profitable, but below that depth the gold could not be extracted profitably.

The *Crump Mine* is situated 2½ miles southeast of the Stewart. The ore has been explored by three shafts for a length of 300 yards. The vein matter consists of quartz in the slate with little disseminated pyrite. The highest assay showed $6.41, which would indicate a very low-grade ore. The gold usually occurred in pockets and nuggets.

The *Fox Hill Mine* is 1¼ miles northeast of Indian Trail. The country rock is a slate, which strikes nearly north and south and dips slightly eastward. The vein, as shown by the 40-foot shaft, was from 6 to 10 inches in width, and consisted of quartz, brown ore, and pyrite.

The *Phifer Mine* is ¼ mile north of Indian Trail. The

country rock is a soft schist, occasionally silicified, which strikes northeast-southwest. The vein consists principally of quartz with some sulphides, but free gold is occasionally found in the quartz.

In the summer of 1933 a small exploration shaft was sunk to a depth of 50 feet. This shaft encountered a pocket of highly oxidized ore which showed high values. Due to the carving in this shaft, the old shaft just south of the vein was collared and cleaned out to a depth of 80 feet. A short drift was sent into the vein and the old workings explored. Careful sampling was done on the quartz vein as well as the country rock, but the values were too low to be worked profitably. Other than the rich pocket of ore encountered in the bottom of the 50-foot shaft, there was no ore in sight which would justify further exploration work. The property was soon abandoned.

The *Black Mine* is ½ mile east of Indian Trail. The country rock is a schist, which strikes a little east of north and dips slightly eastward. There are apparently two veins. The vein, which could be easily examined, varied from 10 to 30 inches in width, but is reported at times to be 4 feet, with a rich seam 10 inches in thickness down to a mere seam. The rich seam showed values of $10.78, $63.30, $101.94, $117.88, and $169.17. The galena showed values between $45 and $50 per ton. The old shaft is only 60 feet in depth, and a drift has been sent out on the vein 30 feet to the north and 40 feet to the south.

The *Smart Mine* is 1 mile northeast of the Black. The country rock is a sericitic schist, at times highly silicified, which strikes N. 30° E., and stands nearly vertically. The vein is quartz with pyrite and galena. The ore body has been proven to a depth of 110 feet, and 3 levels at 35, 60, and 95 feet, respectively, have been run to a length of 60 feet. The vein is 3 to 4 feet in thickness, but the paystreak is much less. The assays show $6.34, $15.51, and $52.46.

The *Secrest Mine* is ¼ mile northeast of the Smart. The country rock is argillaceous to chloritic schist, which strikes N. 30° E. and dips slightly westward. The assays from the vein show $18.80 and $72.22 per ton. The paystreak is an 8-inch seam of brown ore. No reliable information is available relative to this mine.

The *Moore Hill* group of mines is located two miles south of Indian Trail. This group includes Moore Hill, Folger Hill, Davis, Phifer, Lewis, Hemby, and Harkness. Because of their close relationship, these mines are discussed as a group rather than individual mines.

These mines are located in a zone of auriferous schists 3 miles in length and approximately ½ mile in width. As shown by the old workings these mines have been worked down to water level, which seldom exceeds 50 feet in depth. The mineralized zones were capable of being worked across the entire width at times, while at others the miners followed the rich streaks. However, occasional panning shows values 100 feet in width.

The vast number of diggings on the surface is the result of petty leases, and bears witness to the value of the deposits. It is reported that many of the shallow pits produced a large quantity of gold. While none of the mines were systematically operated, yet a great deal of profit was reported from the past operations. The Mint Hill open cut, probably the largest opening in the belt, has been worked to a depth of 50 feet and 100 feet in diameter.

The country rock is an argillaceous schist, which strikes N. 20° to 25° E. standing almost vertically, but dips slightly in an easterly direction at some places and a westerly direction in others, the latter prevailing. The character of these schists varies from very soft to highly silicified types. They are impregnated with small pyrite crystals, at times from 1 to 2 per cent of the rock, and numerous small lenses of pyrite and calcite occur. Granite stringers are also frequent. The schists are occasionally cut by diabasic dikes; especially do these dikes occur at the Phifer, Hemby, and Harkness mines.

The gold is usually not uniformly distributed through the rock, but is found mostly in narrow parallel seams varying in thickness from very narrow to an inch or more. These seams are quite porous and discolored with the oxides of iron and manganese, especially above water level. The ore bodies are from 1 to 6 feet in width, and the barren ground is that portion in which the quartz veins are less frequent.

The soft decomposed nature of the schist above water level permitted an easy and cheap method of mining, but below the water level the mining costs were much greater as the ores changed to sulphides. On account of the high cost of mining and milling the sulphides, these ore bodies are still largely intact.

The *Moore Hill Mine* has been worked to a depth of 70 feet and 100 feet in length.

The *Davis Mine* has been worked to a maximum depth of 150 feet at the road shaft, and fully 500 feet in length.

The *Folger Hill Mine* was worked to a shallow depth but from 300 to 400 feet in length. Some of the rejected ore from this mine assayed $4.35 per ton.

The *Phifer Mine* was worked to a depth of 100 feet and for a length of 400 feet. Some very rich stringers were found in this mine, and at times these stringers were so close together that the whole material could be worked for 100 feet in width and to a depth of 80 feet.

The *Ore Hill Mine* was worked to a depth of 80 feet.

The *Lewis Mine* was worked to a depth of 80 feet and nearly 1,000 feet in length.

The *Hemby Mine* was worked to a shallow depth, but

two shafts have been sunk to 100 feet in depth. The ore body was over 1,500 feet in length. The ore encountered, as shown by the old dumps, contained much solid glassy quartz, as well as some siderite. The ore showed values from $3.31 to $16.57 per ton.

The *Harkness Mine* is reported to have been worked 120 feet in depth and 300 feet along the strike. The ore was rather quartzose which contained much coarse gold of relatively high value.

In 1933 some exploration work was undertaken in this group of mines. At one locality a shaft was sunk to a depth of approximately 40 feet. The ore from this shaft was a soft decomposed schist containing numerous stringers of quartz. An old Lane mill was reconditioned and put into operation to mill the ores taken from the shaft. After several unsuccessful attempts to mill the ore from the shaft, several prospect pits were dug to locate ore which would be amenable to this milling process. Either the ores were too low grade or the recovery was not satisfactory, as the mill operated only for a short while.

The *Bonnie Belle* or *Washington Mine* is located 1½ miles north of Potter's station.

The country rock is an argillaceous schist, varying in hardness from soft to a highly silicified material. It is quite similar to the country rock of the Moore Hill group. The strike is N. 55° E., and the dip at a steep angle westward. The schist shows banding both with and across the schistosity, pointing to their sedimentary origin. A diabase dike intersects the country rock nearly at right angles, and is reported to be 10 feet in thickness. The schists are impregnated with finely divided sulphides, chiefly pyrite with little chalcopyrite. Some free gold occurs as coatings on the major cleavage planes. The maximum width of the ore-bearing zone is stated to be 14 feet, and has been proven over

Fig. 15. Mr. Mack Cox, who Found Eight-Pound Nugget at Reed Mine, 1896.

Fig. 14. Head Frame, Howie Mine, Union County.

a distance of at least ¼ mile along the strike. There are also indications of parallel ore bodies, but these were never prospected. The schists are intersected by numerous quartz veinlets which trend in various directions.

In 1894 a shaft about 20 feet in depth was sunk on the ore body, and it is reported that the ore was from 5 to 8 feet in thickness with a 2-foot richer portion lying on the hanging wall side. A small Chilean mill, in combination with four drag mills, treated the ore. The discharge from the mills passed over amalgamated copper plates, and thence to a Gilpin County bumping table for concentration. The mill had a capacity of ten tons per day. A sample of the ore gave $5.55, and the concentrates showed $22.61.

The *Howie Mine,* which was formerly operated by the Colossus Gold Mining and Milling Company, was purchased by the Howie Mining Company, and extensive preparations were made to develop and operate this property on a large scale. It is located about 3 miles east of the town of Waxhaw, Union County, and 22 miles south of Charlotte. The company owns about 250 acres of ground.

The country rock of this section is a dark-colored argillaceous slate, which is usually very highly silicified, and has a strike of N. 55° E. Traversing this area are many so-called veins or ore seams or streams varying from 18 inches to 16 feet in width. The ore is very similar to that occurring in many other sections of the silicified slates, and it is often very difficult to determine where the vein material ends, and it is usually necessary to constantly test the vein material and limit the mining at that point where the value is so low that it does not pay to take out the ore. It is, however, very rarely that this can be determined by the external appearance of the ore, and assaying is necessary. Pyrite occurs in the vein material and also in the wall rocks. These quartz seams and veins have a strike of N. 55° to 65° E. and a very steep dip from almost perpendicular to slightly to the northwest. Cutting the slates and the quartz seams are several diabase dikes, which have a northwest strike. The rich ore seems to be in the vicinity of these dikes. The ore body might be said to be that portion of the slates adjoining the quartz seams where the gold occurs as fine films on the cleavages and joint planes of the slates.

In the early history of the mine, it was developed by a series of shafts, the deepest of which was the Cureton, 300 feet deep. Most of the mining carried on by the Howie Company has been done by the "Main Shaft" (Cureton). The total depth of this shaft is now 355 feet.

*Two levels have been opened, one at 150 feet and one at 265 feet. At the 150-foot level, drifts have been run southwest and northeast on the vein without any pay ore having been exposed. Cross-cuts extend southeast and northwest from the shaft, also in waste. The total amount of drifting and cross-cutting is about 200 feet.

At the second level, drifts have also been run southwest and northeast, and cross-cuts same as on the first level. In the southwest drift an ore body was encountered about 100 feet from the shaft, and a stope has been started on the ore. Ore is being broken here from 6 to 8 feet wide, averaging by grab samples from $10 to $300 per ton. The higher values have been recently found in the southeast wall of the stope, the rich portion of the vein being 1 foot to 18 inches wide, and averaging from $70 to $400 per ton in gold. Very little silver is found. This rich ore is at present being sacked for shipment, the balance going to the mill.

The mill ore will average from $10 to $30 per ton. The bulk of the gold occurs on the cleavage planes in a very fine state of division, the remainder being disseminated through the quartz in coarser particles. Gold is readily seen with the naked eye in the rich ore.

The bottom of the shaft, 90 feet below the second level, shows assay values of $2.40 per ton, and the intention is to sink further, with a view of cutting the same ore shoot now being worked on the second level.

This ore shoot is south and east of the old workings in the Bull Face shaft, operated some years ago by the Colossus Mining Company, and sunk on an incline to a point 90 feet below our second level.

There is a break through from the southwest end of our drift on the second level into these old workings, but up to the present time water has prevented any examination of the old work, and their extent has not been ascertained by the present company. According to report, a consid-

*From letter of Robert Clark, superintendent of the Howie Mine, to the State Geologist, December 26, 1913.

erable amount of ore has been mined from these old workings. The Bull
Face shaft is vertical for 150 feet from the surface, the remainder of the
distance being on an incline following the dip of the ore shoot. There is
said to be good ore left in the bottom.

Northwest from the Bull Face are two shafts, the old and new
Neddy shafts. These are in bad condition, and are not being used. The
new Neddy produced a considerable amount of good ore from the Neddy
vein to a depth of 300 feet. This vein parallels the Bull Face vein on the
northwest. There is a cross-cut connecting the Bull Face and the new
Neddy shafts at 150 feet.

At a point 1,500 feet northeast from the main shaft the company
is developing a vein exposed in some old workings. This shaft is 60 feet
deep and has been equipped with a small headframe and hoist. A 1½-
inch air line from the main shaft supplies compressed air to operate the
hoist and machine drills. Sufficient development work has not yet been
done in this shaft to determine the extent of the ore shoot, but good ore
is being broken in an underhand stope from 2 to 3 feet wide.

Ore is trammed from the main shaft to storage bins, from which it
is taken to the mill by an inclined tramway. This incline is to be extended
on a 45-degree dip to cross-cut the vein systems in this the western end
of the property.

The *Old Colossus* or *Howie Mine,* near Waxhaw, the
principal mine in Union County, was actively developed for
five months in 1913, and, it is reported, was opened by a 355-
foot vertical shaft and a 100-foot adit. The property was
equipped with a 50-ton all-sliming cyanide plant, in which
continuous agitation and decantation were practiced. The
ore is sulphide with siliceous gangue and is oxidized to a
depth of 40 feet. In 1914 the greater portion of the gold
production from this county came from this mine. The mill
was operated intermittently in 1914 for about four months.
In 1915 the Howie was also the largest producer in Union
County. In November of 1915 a 300-ton cyanide plant was
built at this mine. It uses the Dorr system, which is said
to have made an excellent recovery for the siliceous gold ore.
In 1916, however, fire destroyed the shaft house of the
Howie Mine, which was not repaired until December, 1916.
This made it necessary to unwater the mine, construct con-
crete retaining walls and piers to support the permanent

head frame, and on March 12, 1917, the secretary of the company reported that they had, at the time of writing, almost completed the wall. He states: "Our mill will start within the next six weeks on a 50-ton daily basis, and we resume mining operations this week, using a temporary head frame and sinking winze on the Bull Face ore body at the third level; and following on the ore chute at an incline. Development work will also be pushed, as well as the dragging of ore from other exposed bodies on each of the levels. The Bull Face ore, which runs from $40 up, will be shipped to smelter until we have installed concentrators."

In 1935 the Condor Consolidated Mines, Ltd., Toronto, Canada, had control of the Howie Mine. This company employed the Milton Hersey Company, of Toronto, Canada, to give a report on the property. The old shafts and drifts were unwatered, and a series of 48 core drill holes were made. The results of this investigation revealed approximately 70,000 tons of ore, which averaged about $15 per ton. As a result, plans are under way for the construction of a 100-ton cyanide mill.

ANSON COUNTY

In Anson County, about 2 miles southwest of Wadesboro, there is a small patch of crystalline rocks lying on the south side of the Triassic sandstone belt, which is gold-bearing. Two mines have been worked to some extent in this area.

The *Hamilton* or *Bailey Mine* is situated 2 miles southeast of Wadesboro. There are two quartz veins, one of which was worked to a depth of 100 feet, said to be from $2\frac{1}{4}$ to 4 feet in thickness. The assays showed values from $4 to $30 per ton. In 1934 some exploration work was carried on at this property, but nothing of value was uncovered.

The *Jesse Cox Mine* is located in this same vicinity.

CHAPTER VI

THE CAROLINA IGNEOUS BELT

GENERAL STATEMENT

The Carolina Igneous Belt is the second most important gold-bearing section in North Carolina. This belt is bounded on the east by the Carolina Slate Belt and on the west by the gneissic rocks of the Piedmont section. It extends from the northern section of Caswell and Person counties in a southwest direction into the northern part of South Carolina. It varies in width from 15 to 30 miles, and covers parts of Caswell, Person, Alamance, Guilford, Randolph, Davidson, Davie, Rowan, Cabarrus, Mecklenburg, and the eastern edge of Lincoln and Gaston counties. The gold deposits are confined largely to Guilford, Davidson, Rowan, and Mecklenburg counties. Mecklenburg County is by far the most important gold-bearing section of this belt. The total gold-bearing area of this belt is approximately 1,000 square miles.

The rocks of this belt, as inferred by the name, are principally igneous, especially granites, with basic intrusions as diorite, gabbro, and diabase. At times it is impossible to classify the rocks, as there is a gradual transition from one into the other. In certain areas of this belt the rocks are highly metamorphosed, and a schistose or gneissic structure has been developed; especially is this true near Lexington in Davidson County and at the Rudisil Mine in Mecklenburg County.

The rocks of the Igneous Belt are younger than those of the Slate Belt as shown by the nature of the contact. This is especially shown in Union and Cabarrus counties.

Large granitic and diabasic intrusions intersect the main igneous mass, which would indicate that these intrusions are still younger than the rocks of the main belt. These large intrusions were apparently intruded before the force producing the schistosity was active. Near the center these large intrusions are coarsely crystalline, while near the borders they are finer grained.

The quartz veins found in the Carolina Igneous Belt are the fillings of fissures. The fissures are quite regular in outline as at the Reed, Phoenix, and Randall mines. At times they appear as pockets or lenticular bodies along the contact of diabase dikes with the granite. Occasionally the quartz veins are confined to the diabasic intrusions.

The gold ores in the Igneous Belt were deposited by ascending mineralized water. The ores carry, in addition to the gold, varying amounts of copper, silver, lead, zinc, and occasionally nickel. Arsenic and antimony are not common. All of the ores are refractory at depth only as the sulphides make them so.

The chief gangue mineral filling the fissures is quartz carrying gold-bearing sulphides, especially pyrite and chalcopyrite. In a great many of the mines in this belt, near the surface there are finely laminated, argillaceous or chloritic slates; at depth, however, this slaty structure disappears. The slaty structure is sometimes included within the quartz; generally, however, it occurs nearer the wall, grading gradually into the country rock. The quartz veins intruding the slaty structure may be classed as injected vein dikes. Sometimes two or three quartz veins are noticed with slate partings, but usually evidence goes to show that they are all parts of the same vein. Near the wall the quartz is frequently banded.

Many of the mines in this belt were worked as gold mines near the surface, but later, as copper was encountered at depth, developed into copper mines. The *Fentress Mine* in Guilford County is a good example of this type. The mines of the Carolina Igneous Belt are concentrated in the southern end of Guilford, the eastern section of Davidson, Rowan,

Cabarrus, and in the central part of Mecklenburg in the vicinity of Charlotte. The most important mines are located in Cabarrus and Mecklenburg counties.

GUILFORD COUNTY

The mines in Guilford County are located in the southern portion south of Greensboro.

The *Hodges Hill Mine* is situated 6 miles southeast from Greensboro, and lies near the eastern border of the Granite Belt. The vein varies in width from 6 inches to 12 feet, and has been opened up by pits along the outcrop for a distance of 800 or 900 feet. The vein is essentially quartz lying rather flat, and contains other gangue minerals, as pyrite, siderite, chalcopyrite, and limonite. The chalcopyrite at times has been altered to malachite and red oxide. Assays showed $1.03, $2.07, $11.36, $22.74, and $46.05.

The *Fisher Hill* and *Millis Hill mines* are approximately 6 miles slightly west of south from Greensboro. It is reported that fifteen veins are found on the mining property. There are two systems of veins, one which runs approximately north and south, and the second nearly northeast and southwest. The dip of the veins varies from 15° to 20°. The veins are composed of quartz carrying heavy percentages of pyrite with little copper.

The veins have been traced for a distance not less than 8 miles, though not capable of being worked the entire distance. However, the vein has been successfully operated at several points in the 8 miles distance. The ore body varies from 4 inches to 10 feet in thickness, and at times carries relatively high values.

The *Twin Mine* is 6 miles southwest from Greensboro, and derives its name from the fact that two parallel veins are exposed in one tunnel. The veins strike N. 40° E. and dip southeastwardly. The veins are about 18 inches in thickness with 4 feet of slate between. They are composed of quartz and carry small amounts of chalcopyrite.

The *North Carolina* or *Fentress Mine* is located 9 miles south of Greensboro, near the eastern edge of the Granite Belt. The old mine dumps bear witness to the size of the vein as well as to its length, for it has been traced 3 miles along the outcrop. However, the most active development did not exceed ½ mile in length. The vein has been worked by three deep shafts, the deepest of which is about 400 feet. The shaft is partly vertical, about 330 feet, and the remainder on the incline. Four levels have been run from 300 to 500 feet in length from the main shaft.

In the extreme southwest end of the vein the ore body is 3 to 4 feet in thickness at 40 feet in depth. At 310 feet the fissure is 7 to 13 feet wide between the walls. There was no improvement, however, in the vein where it expanded, and it carried no more copper than when 7 to 8 feet thick.

In the beginning this mine was worked as a gold mine, but at depth the copper content ranged from 14 to 23 per cent, therefore was later worked as a copper mine. At times the copper sulphides lay on the foot wall, then suddenly curved upward and followed the hanging wall for a distance of 20 to 30 feet. Assays from the old dumps showed values from 62c to $6.97 per ton.

The *Gardner Hill Mine* is 2 or 3 miles to the northeast of Jamestown. Little work has been done at this mine since 1856.

The country rock is principally granite, but on either side of the vein the ore is bounded by slate or killas. The vein is from 6 inches to 3 feet in width and has been explored to a depth of 258 feet. It is composed of quartz, which is commonly the case in the granite, with pyrite and chalcopyrite. There are said to be three veins on the property, but their relation has not been determined.

The Creek shaft is said to be 110 feet; the Underlay shaft, 600 feet S. W., 175 feet deep on the incline; the Old Engine shaft comes 400 feet from the south end of the property, 175 feet vertical; the New Engine shaft is 258 feet in depth; the No. 2 shaft, 110 feet vertical; and the White Oak shaft, 150 feet vertical. Four levels have been sent out at the 60-, 100-, 150-, and 228-foot depths. They average 500 feet in length, and have been driven in both directions from the shafts. It is reported that the mine is pretty well stoped out from the water level to the bottom of the respective shafts. The ore near the surface was free milling, while at depth it ran into copper sulphides.

When copper ore was encountered in quantity, little effort was made to extract the gold; shipping ores of copper were exclusively sought after, and it is reported that for many years a very high degree of prosperity was enjoyed. At the height of activity 40 tons of yellow copper ore were shipped a week, averaging from 20 to 25 per cent copper. The width of the veins at times was 20 feet. Assays on the ore showed from $3 to $10 per ton in gold. It has been estimated that 25,000 tons of ore are on the dump.

The *North State* (or *McCullough*) *Mine* is located two miles west of south from Jamestown. Emmons states that the vein fissure pursues a northeast course, but is curved in the middle. It dips S. 80° E. and at one place S. E. The vein is composed of a column of brown ore resting on the footwall, which extends from the outcrop to 130 feet in depth. Upon this rests the disintegrated ore, containing 8 inches of beautiful copper pyrites, and then against the hanging wall, quartz rather poor in gold is frequently 8 feet thick. The vein at the surface is about 2 feet wide; at 60 feet it is 4; at 90 feet it is 10; and at 130 feet it is 24½ feet wide. It dips at an angle of 45°. At the 130-foot level it swells out into a rather lenticular form. Here the ore is concretionary; on the footwall the brown ore is 6 inches thick only, then copper pyrites, then a belt of brown ore containing nodules or concentrates of pyrite more or less changed, the middle of which is rich in gold. Upon the hanging wall is the principal mass of porous quartz, which is generally poor. The brown ore is soft and easily crushed. It is intermixed with fine quartz and spongy masses of it, which are rich in gold. * * * Towards the north is Jacks Hill; a shaft sunk in the top of this hill cuts the vein at a depth of seventy-seven feet, where it is seventeen feet thick. * * * The copper is the purest sulphide, yielding by analysis thirty per cent of copper. * * * The McCullough vein carries its gold in combination with the sulphides.

Later information shows that the shafts reached a depth of 392 feet (about 325 feet vertical). The vein at the bottom of the shaft showed 4 to 8 feet in width, and was composed principally of quartz carrying sulphides, sometimes massive, but generally scattered. The mine was abandoned in 1885.

The *Lindsay Mine* is apparently the southerly extension of the North State. The deeper shaft was 150 feet vertical. A second vein was discovered 22 feet behind the main vein but it was never stoped. A series of 87 assays taken from all parts and from all classes of ore at the mine ranged from $4 to $100 per ton.

DAVIDSON COUNTY

The *Lalor* (or *Allen*) *Mine* is located 2 miles southeast from Thomasville. Three shafts have been sunk, the deepest of which is 165 feet on the incline or about 140 feet vertical. The vein is said to be of good width, and carries a fair proportion of iron and copper pyrites. The mine was operated in 1882, at which time a 10-stamp mill was erected. However, by 1886 it was abandoned.

The *Eureka Mine* is ½ mile west of the Lalor. The deeper shaft is 125 feet in depth. The vein and character of the ore is quite similar to that of the Lalor Mine.   Assays showed values from $7 to $45 per ton.

The *Loftin Mine,* in Davidson County, is 1½ miles southeast of Thomasville.   The nature of the vein is quite similar to that of the Lalor Mine.

ROWAN COUNTY

The *New Discovery Mine* is 3 miles east of Salisbury. The deeper shaft is 100 feet.   Very little is known relative to this mine, as it ceased operations at the close of 1883.

The *Dunns Mountain Mine* is situated 3½ miles east of Salisbury.   It is stated that there are three veins, the first of which was worked to a depth of 190 feet, averaging about 4 feet in width.   This vein was largely filled with slate and quartz, and carried a small proportion of pyrite with traces of chalcopyrite.   The second shaft or Office shaft was worked to a vertical depth of 140 feet.   It is reported that the ores were oxidized nearly to the bottom of the shaft, and therefore contained little sulphides.

The *Reimer Mine* is situated 6 miles southeast of Salisbury, on the waters of the Yadkin River.   The mine was opened by 3 vertical shafts, 193, 43, and 165 feet in depth, respectively.   The vein is principally quartz, carrying about 10 per cent sulphides, varying in width from 1 to 9 feet with an average of about 3½ feet.   The quartz has a banded structure, which more or less parallels the walls.

The last work accomplished at the Reimer Mine was in 1895 in the lower levels of the 193-foot shaft.   Assays on the ore showed from $4 to $24 a ton with an average of $4 or $5.   The work at this mine was terminated in 1883 by the burning of the buildings and the destruction of the machinery.

The *Bullion Mine* is located 1½ miles east of the Reimer. Outcrops of the vein show a good width, but for some reason have never been worked to any great extent. At the 90-foot level a drift was driven 200 feet in length on a vein reported to be 7 feet in width. The last work reported was in 1881. Assays on the ore showed $6.20, $9.32, $10.34, and $15.51 per ton.

The *Dutch Creek Mine* is situated 10 miles southeast of Salisbury. It is reported that several veins are found on this property, which have a northeast-southwest trend. Several shafts have been sunk, which exposed sulphides at water level. Little is known of the width of the veins or the character of the ore, other than that varying percentages of copper are encountered.

The *Gold Knob Mine* is 9 miles southeast of Salisbury. Three veins are reported, the widest of which is in places 20 feet. The ore is rather low grade and carries low percentages of copper. Chimneys of high-grade ore were encountered.

The *Atlas* and *Bame mines* are supposed to be the southwest extensions of the Dutch Creek veins. The veins are reported wide, but as a rule rather low grade, at least as far as free milling qualities go.

MECKLENBURG COUNTY

The *Rudisil* and the *St. Catherine mines* are respectively in the south and the north parts of the same lode; the former is one mile, and the latter one-half mile south of the center of Charlotte.

The bodies of ores in both mines nearer the surface lie in slates, which in places approximate 100 feet in width. These slates are both argillaceous and chloritic, everywhere siliceous, and frequently replaced by layers of quartz with ore. They are bounded by the country rock of the igneous belt, massive crystalline rocks.

At the outset, and to the depth of something more than

100 feet, two bodies of ore were recognized—the "back vein" and the "front vein"—but the intervening mass of slate frequently carried subordinate bodies.   Emmons says of the Rudisil Mine: "The rock both above and below the two veins, which constitute the mine, is the syenitic granite of the Salisbury and Greensboro belt; but the veins are immediately in killas or slate, which cannot be distinguished from the slates which predominate in the slate belt; and there are other points where the slate is in granite, and not less than 100 feet thick, which is traversed with veins of quartz.   It is difficult to determine whether the slate thus situated is to be regarded as the killas of the vein, or as masses of the slate system isolated by an eruptive rock."

"Thus the Rudisil veins are between masses of an eruptive rock. * * * The vein fissure is fifty feet thick, occupied by talcose slate, which is overlaid by white granite, and underlain by elvan or a dark trappean rock."

The strike of the fissure approximates N. 30° E., and the dip 45° N. W.   At the depth of 200 feet or more the slaty character becomes less evident, and ultimately disappears, seemingly giving place to the massive country.

The two ore bodies (front and back veins) vary considerably in width from 2 to 4, and sometimes 6 feet.   At the depth of 200 feet they appear to approach.

The ore was carried in pockets in the slates (or schists), and in great abundance, so that for many years the mine was very prosperous.   At the surface, and to a considerable depth, the mine material was the rich and easily treated brown ore of the region.

The zone below water level carried iron pyrite with a little copper pyrite; and the ore was scattered somewhat through a slaty and quartzose gangue, being less pockety than that above the water line.   The assays for this zone as a rule showed material of only moderate value.   At the

200-foot level the peroxides have mostly disappeared, and with them largely the free gold, though both are present in some proportion to the lowest level—350 feet below the surface.

At greater depths the sulphides were scattered thinly through a quartzose and somewhat slaty gangue, or in narrow seams, or concentrated in large, wide and rich shoots of nearly solid sulphides. An inspection of the plate will show the position of these shoots—3 in number. As regards the northern shoot, neither the point of origin nor its character are matters of record; it is not known above the 130-foot level.

The south and the middle ("Burnt Shaft") shoots began just above the 130-foot level in narrow threads, and expanded both in width and in length at greater depths. Neither the north nor the middle shoots have been followed below the 192-foot level to any extent, though slightly explored from the 250-foot level. The north shoot increased from a mere seam to 5 feet in thickness, and had a length in the direction of the vein, varying from 30 to 50 feet. The material of this shoot was a high-grade sulphide, but not the best that the mine has furnished. The middle shoot increased in the same way to be 8 feet thick, extending longitudinally from 10 to 50 feet; the grade of ore is believed to have been somewhat higher than that of the north shoot.

It was, however, to the Big shoot (South shoot) that the mine has in recent years (as late as 1887) owed its reputation. It commenced like the middle shoot almost in a point, and gradually widened and lengthened with a slight south pitch of its own in the vein, till it became in places 15 feet thick, and 100 feet long; its ends were not abruptly marked off from the adjacent vein or from the "country," but passed gradually into mine material comparatively barren, or into "country" quite worthless.

The contents of the shoots, as a whole, were compact iron

pyrite with a very little copper pyrite, and some quartz, which later was, however, for the most part readily cobbed out. This ore was uniformly of high grade, entire shipments sometimes ranging as high as $180 per ton.

This shoot extended down a little below the 300-foot level, but in the 350-foot level it has never been found. There has been much speculation about it, and opinions have varied as to whether it had disappeared altogether, or was simply "thrown" from its normal position by one of the many deflections of the vein from its direct course.

It may be added in conclusion that the so-called barren parts of the vein carry pyrites scattered through the quartz and slate, and most of it is quite capable of being treated by a preliminary cobbing to separate the massive pyrites for direct metallurgical treatment, and at the same time prepare the lowest grade material for battery amalgamation and subsequent concentration preparatory to smelting or chlorination.

The following assays show the character of some material from this mine:

ASSAYS, ORES AND CONCENTRATES, RUDISIL MINE, MECKLENBURG COUNTY.

Assays of Auriferous Pyrite Scattered in Gangue.

Gold, per ton	$6.20	$ 9.30	$12.40	$20.67
Silver, per ton	Trace	2.49	Trace	.13
	$6.20	$11.79	$12.40	$20.80

Assays of First Class Ore, Rudisil Mine.

Gold, per ton	$45.47	$73.35	$74.41
Silver, per ton	Trace	Trace	Trace
	$45.47	$73.35	$74.41

Assays of Second Class Ore, Rudisil Mine.

Gold, per ton	$24.80	$28.70	$29.97	$31.01	$35.14	$35.14	$36.18
Silver, per ton	Trace	Trace	.19	Trace	.65	.96	.13
	$24.80	$28.70	$30.16	$31.01	$35.79	$36.10	$36.31

Assays of First Class Cobbed Sulphurets, Rudisil Mine.

Gold, per ton	$165.36	$126.09	$227.37
Silver, per ton	.45	.94	1.71
	$165.81	$127.03	$229.08

Assays of Concentrates, Rudisil Mine

Gold, per ton	$56.64	$59.94	$67.18
Silver, per ton	2.90	1.73	1.14
	$59.54	$61.67	$68.32

In 1934 the Rudisil Mine was unwatered and further explorations made. The results were so favorable that the Rudisil Mining Corporation was formed to develop the mine on a large scale. By the middle of the year 1935 approximately $75,000 were spent in developing the mine, the erection of flotation mill, and in the erection and equipping of a laboratory.

The management reported that approximately $40,000 worth of concentrates were shipped during the months of June, July, August, 1935.

Exploration work revealed three new veins, which averaged from 3 to 5 feet in thickness with the gold value varying from $5 to $75 per ton with an average between $15 and $20 per ton.

The concentrates from the flotation mill are shipped to the American Metals Company, Carteret, New Jersey, and are said to average from $100 to $125 per ton. Due to the freight and smelter charges, which average about $12 per ton, some thought was given to the erection of a small smelter with a capacity of 20 to 25 tons per day. However, nothing definite had been decided by November 1, 1935.

The *St. Catherine (Charlotte) Mine* is located in the northeast extension of the Rudisil lode, the two mines being over half a mile apart. The region intervening has been prospected superficially, but as yet nothing of importance has been found which promised well for deeper operations. Both mines have the same general features, and agree in strike and dip.

The St. Catherine has been worked to a depth of 460 feet (155 feet vertical and 305 feet on the underlay, equivalent to a total vertical depth of 370 feet).

In the lower part of the mine the vein does not seem to be well consolidated, and the geological relations are perplexing. Below 250 feet there are several large shoots of low-grade ore quite suitable for milling and concentrating, notably the ore body worked from the "pump" shaft and between the depths of 200 and 370 feet, below which point it has not yet been removed. The occurrence may be briefly stated as a series of obscurely parallel seams of slate, with quartzose ore bodies 2 to 6 feet in thickness between; the amount of pyrite in this class of ore is small.

A cross vein (striking N. W. and S. E.) has been examined from the 155-foot level for a distance of 100 feet along the vein.

The following assays show the range in character and value of some of the ores of this mine:

Assays, Gold Ores, St. Catherine Mine, Mecklenburg County.

Assays of Brown Ores.

Gold, per ton	$26.87	$39.27	$56.87	$103.35
Silver, per ton	.71	.58	.38	.88
	$27.58	$39.85	$57.25	$104.23

Assay of Quartz, with Disseminated Pyrite, St. Catherine Mine.

Gold, per ton	$24.80
Silver, per ton	.16
	$24.96

Assays of First Class Ores, St. Catherine Mine.

Gold, per ton	$52.19	$53.74	$72.41
Silver, per ton	.55	Trace	.39
	$52.74	$53.74	$72.80

Assays of Second Class Ores, St. Catherine Mine.

Gold, per ton	$35.14	$33.07	$35.14
Silver, per ton	1.14	Trace	.28
	$36.28	$33.07	$35.42

Assays of Cobbed First Class Ore, St. Catherine Mine.

Gold, per ton	$95.08	$108.52	$181.38
Silver, per ton	.41	Trace	Trace
	$95.49	$108.52	$181.38

Assays of Concentrates, St. Catherine Mine.

Gold, per ton	$40.31	$66.14	$133.32
Silver, per ton	1.10	1.23	1.23
	$41.41	$67.37	$134.55

In 1883 a 10-stamp mill was erected. The ores were first subjected to a preliminary cobbing, which separated out the massive pyrites and lean ore, the latter going to the stamp mill. The free gold was caught in the battery and on the plates in the customary mode of amalgamation, and the tailings led directly to Frue vanners, where the product was concentrated. The cobbed pyrites and concentrates were shipped north and elsewhere for treatment.

The proportion of massive pyrites to the whole material mined (except when near a rich chimney) was small, probably not more than 2 or 3 per cent.

In the common run of mining and milling practice it required 10 to 15 tons of ordinary ore to make one ton of concentrates, which ordinarily contained 80 to 90 per cent sulphides. It is worthy of remark that the concentrates, however high the per cent of sulphides, rarely contained as much gold per ton as cobbed ore of the same richness in sulphides.

The last work at the St. Catherine was in 1887.

The *Smith* and *Palmer Mine* is one mile south of Charlotte, near the Rudisil. Some believe that this mine is an extension of the Rudisil vein, or at least a parallel body. A line of pits indicate vein matter for a distance of 500 feet along the strike. The greatest depth of the workings was 75 feet, and the width of the vein varied from 2 to 4 feet. Assays showed values of $5.17, $5.46, $15.51, $15.77, and $149.73 per ton.

The *Howell Mine* is believed to be on the southern extension of the Rudisil vein. It has been worked to a depth of 32 feet, and approximately 50 feet in levels have been driven. The vein is stated to be from 2 to 4 feet wide with assays showing from $5 to $14 per ton on the brown ores, and from $38 to $77 per ton in the sulphides.

The *Clark Mine* is 2½ miles west of Charlotte. There are apparently two vein systems, one northeast and southwest, and the other nearly east and west. The northeast and southwest system was worked to a depth of 70 feet, as shown by the old workings, for a distance of about 1,200 feet. It is reported that the mine was abandoned on account of flooding by water. Assays on the vein showed $5.17, $32.22, and $147.05 per ton.

The east and west vein was worked to a depth of 78 feet. At the 72-foot level three boxes of low-grade brown ore were found within a space of little less than 25 feet. Only one of the bodies of ore discovered, however, was rich enough to work; it averaged $16 per ton.

The *Stephen Wilson Mine* is 9 miles west of Charlotte. Ten well-defined veins are reported, only two of which have been worked to any extent. The No. 2 vein is from 2 to 3 feet wide, and was worked to a depth of 400 feet on the incline. Low percentages of copper were reported. Assays on the ore showed $1.09, $25.70, $52.10, $97.54, $156.58, and $345.96 per ton.

The *Capps Mine* is situated 5½ miles northwest of Charlotte, between the Rozzel's Ferry and Beattie's Ford roads.

It is located on a group of veins of which two are closely convergent (the Jane and the Capps). By the accident of different ownerships, they have been for the most part separately and differently developed. The Capps vein has a strike N. 30° to 35° W., and a dip westerly, with some variations of about 40°. The Jane vein runs N. 40° to N. 60° E., and has a very steep pitch eastward. It is certain that the actual intersection of these veins has been found. The Capps vein has an ascertained length of nearly 3,000 feet, and the Jane probably fully as much.

As the development of these two veins has been separate, the description will follow the course of the work.

The later work on the Capps was restricted, and finally stopped, from legal considerations, but the earlier work extended very nearly along the line of the outcrop of the entire vein—2,000 to 3,000 feet—and was carried to such depths as to disclose clearly the character of the outcrop. At points where the ore proved to be exceptionally abundant and valuable, operations were extended much deeper. There is perhaps no vein in the whole section which shows such extensive prospecting on the surface, and bears all the appearance of having been highly remunerative.

It is much to be regretted that there are such scanty records of the earlier work, and of the characteristics of the veins and deposits. The outcrop of the vein at many points still shows a width of 20 to 25 feet, and the debris everywhere indicates a very wide vein.

The ores near the surface are the customary brown ores with quartz, and are generally free milling.

They were not uniformly disseminated in the quartz, but generally occurred in layers, sometimes near the hanging wall, sometimes near the foot wall. At greater depths sulphurets of iron (and to a small extent, of copper) together with quartz, were found; nevertheless, at the depth of 130 feet there was still much brown ore.

The work was never prosecuted to any great depth—at the Gooch shaft 70 feet, at the Bissell 130, and at the Penman 65 feet.

The filling of this vein is quartz; its width is not known, for no systematic work has ever been undertaken to find the walls; it cannot be less than 20 feet, and possibly is considerably more.

The line separating the veins from the walls is not always sharp and definite, and occasionally, where the supposed hanging wall had been reached, another and valuable body of ore was found still further beyond, which ultimately came back to the main vein.

There are also "cross courses" leading into the main vein nearly at right angles. One of the most conspicuous and valuable of these was found in the 130-foot level, at the distance of 225 feet southeast of the Bissell, and about one-half way between the Bissell and Mauney shafts. It received temporarily the name of the "East and West vein," for want of exact data as to its relations to the main vein. This body departs abruptly from the main vein easterly and towards the Jane vein, and has been followed in that course 120 feet, and almost every foot of the vein material was ore. The width of this deposit could not have been much less than 18 to 24 inches, and the ore was of more than average value.

A few feet further north there is a similar cross course or body of ore, but, so far as explored, it was not so valuable as the "East and West vein."

The Capps Mine has been noted for the amount of ore it could furnish, and for the superior grade of its ore. There are four well known bodies of ores. The first of these is near the Gooch shaft, toward the south end of the mine, and from the 78-foot level downward it yielded largely brown ore with some sulphides, and seemed to improve in character going south.

The second body is in the 78-foot level from the Mauney shaft. Not much can be said with precision respecting the value of this ore, though the upper part of the body toward the surface yielded an ore of high grade.

A third and very large ore body was worked out through the Bissell shaft to the depth of 90 feet. The entire length of this level is 300 feet, of which 200 feet are to the north, and 100 feet to the south of the shaft; the whole of this distance was ore, free-milling to a great extent above, but more and more sulphuretted below; it has never yet been entirely extracted above the 90-foot level, but the best part of it has been stoped out. Below the 90-foot level the ore has not been stoped out at all, except as it was necessary to remove it in running the levels. The body has been explored by a few winzes run downward toward the 130-foot level. Its connection with the large body developed in the 130-foot level has not yet been established conclusively, but there seemed to be little doubt from its position and character that there is such a connection. This body in the 130-foot level is found at a point 125 feet south of the Bissell shaft, and extends north as far as the work has been prosecuted. The shoot cannot be less than 200 feet long, and judging from the 90-foot level, it may be 300 feet.

Assays, Gold Ores, Third Ore Body, Capps Mine, Mecklenburg County.

Gold, per ton	$11.72	$15.51	$17.92	$25.84	$49.61	$132.29
Silver, per ton	Trace	.32	.14	.13	.97	.84
	$11.72	$15.83	$18.06	$25.97	$50.58	$133.13

A fourth body was found in the bottom of the Penman shaft, 335 feet north of the Bissell. The stopings north of the inclined shaft are very extensive and reach to its bottom; the ore body could not have been less than 3 feet thick on an average, and this increased in one place to be 8 feet; a very little good ore is still to be seen at the bottom.

The deposit in this part of the mine is comparatively shallow, and is, and will continue for some distance, free-milling. The facts recited will justify the expectation of large and valuable bodies of ore at still greater depths.

That part of the Jane vein on the Capps mining tract was worked in part from the Isabella shaft to a depth of 160 feet. There is no record of the value of the ore body at this point; common report speaks well of it, but admits the refractory character of the ore.

The Capps Mine was reopened in 1882, and during the following year some ore was shipped to the Designolle works, 4 miles south of Charlotte. However, this process of reduction was unsatisfactory. In 1884 a 10-stamp mill was erected at the mine, and the ores from the dump were milled; shortly after that time all operations ceased. In the spring of 1895 four diamond drill holes were bored on the Capps vein, respectively 340, 255, 180, and 170 feet deep. The vein was cut by each hole, and showed a thickness of 20 feet, with a dip of about 30° S. W. The walls were fine and coarse grained diorite, at times porphyritic. Assays of the vein matter from the drill cores gave $6 to $7 per ton.

The *McGinn Mine* adjoins the Capps on the north.

The *Jane vein* has been worked from various points, and especially from the Engine shaft (150 feet), at which point all of the readily accessible ore bodies above the 150-foot level have been extracted.

The *Copper vein* was operated extensively for copper to a depth of 110 feet; i.e., as far down as could readily be done some 50 years ago, with the appliances then at command. The ore was yellow sulphide, and was shipped from the State for treatment.

The following assays are appended:

Assays, Gold Ores, Jane Vein, McGinn Mine, Mecklenburg County.

Gold, per ton	$2.07	$5.17	$23.25	$28.42	$99.22	$108.52	$113.68
Silver, per ton	1.20	1.74	.31	.31	.71	1.65	.77
	$3.27	$6.91	$23.56	$28.73	$99.93	$110.17	$114.45

Copper, per cent   8.05%--4.55%

Assays, Ores from Copper Vein, McGinn Mine, Mecklenburg County.

Gold, per ton	$5.17	$10.85	$12.40
Silver, per ton	1.74	1.65	1.20
	$6.91	$12.50	$13.60
Copper, per cent	7.50%	4.55%	8.05%

The *Means Mine* is situated one-half mile southeast from the Capps. Some believe that this mine is a continuation of the Jane vein of the Capps Mine, while others believe it to be an extension of the Capps vein. It is doubtful, however, that either of these opinions is correct.

The vein has been worked at different points and in one shaft, the Wallace, to a depth of 175 feet. The exposure in the face of the drift showed 42 inches of quartz with 16 inches of schist between. The ore from this mine carried considerable chalcopyrite. Assays showed a gold content of $8.67 and $30.38 per ton.

The *Hopewell* (or *Kerns*) *Mine* is 11 miles northwest of Charlotte. It is reported to have had a high yield in former years. The last work was done at the 140-foot level in a vein which showed good yellow copper ore about 2 feet in thickness. Assays on the ore showed $4.13, $12.40, $17.63 per ton in gold, and from 12 to 18 per cent copper.

The *Chapman* (or *Alexander*) *Mine* is 8 miles northwest of Charlotte. It is reported to have been worked to a depth of 110 feet, and that at the 90-foot level, drifts 75 feet were sent out on the vein in both directions.

The length of the vein is said to be 900 feet, the strike N. 20° W., and the dip 65° to 70° N. E. The vein matter is made up of silicified schists with seams of brown ore and quartz. The width of the vein reaches from 9 to 10 feet with an average of much less. The ore seams vary from 4 inches to $2\frac{1}{2}$ feet, and show values from $12 to $28 per ton. The sulphides, however, gave $48 per ton. Other assays showed $4.13, $26.67, $24.03, $24.87, and $35.88 per ton.

The *Dunn Mine* is 2 miles northwest from the Alexander, towards Rozzel's ferry, on the west side of Long Creek.

It was the first mine discovered in the county, not long after the finding of the nuggets at the Reed Mine in Cabarrus County, in the first years of the past century.

The "East vein" was the first to be mined, but the ore soon changed to copper pyrites, which, though auriferous, proved to be too refractory to be treated by the methods then in vogue, and the vein was abandoned after working to a depth of 20 feet. Above this point the ores were mostly peroxidized. The vein is 6 to 12 inches wide; its course is nearly north and south, and its dip west.

The deposit which has been most largely worked is known as the Main vein—a body of slates bearing northeast and southwest, and dipping southeast at an angle of nearly 45°. This body of slates extends across the property for a distance of $\frac{1}{2}$ mile along its outcrop. It contains deposits of quartz and brown ores (including a very hard red hematite, more nearly resembling specular iron), cellular quartzose ores, and compact pyrites, including some copper pyrite. Another "vein" is found 50 feet back of this, but the sections indicate that the whole is one body, with a front and a back seam of ore, and not properly two veins. The appearance at the 60-foot level, where three bodies of ore are seen within a few feet of each other, gives strength to this view, that they are subordinate seams of the same vein, and makes it a not unreasonable supposition that they may combine in depth to form one ore body.

The underground work consists of a shaft 60 feet deep, and a second shaft of 90 feet, which is considerably to the east of the veins; a cross-cut has been driven from the bottom of this shaft to meet the veins. The 60-foot shaft is connected with the 90-foot shaft by a level (the 60-foot level) across the formation. Three, if not four, parallel bodies of ore are cut across by this level. These bodies are composed of silicified slates, varying from 3 to 5 feet each in thickness, and with an aggregate thickness of not less than 12 feet. No drifting had been done on these parallel bodies, and their character and strength are not known.

The following assays show the variation of different samples of ore:

Assays, Gold Ores, Dunn Mine, Mecklenburg County.

Gold, per ton	$8.27	$10.33	$28.94	$128.44	$26.17
Silver, per ton	Trace	Trace	1.94	Trace	Trace
	$8.27	$10.33	$30.88	$128.44	$26.17

The *Henderson Mine* is situated 7 miles north of Charlotte. From the workings the vein appears to be about 300 feet in length. The deepest shaft is 100 feet, and three bodies of ore were worked from it, varying from 1½ to 4 feet in thickness. At the bottom of the shaft sulphides predominate, but brown ores were not entirely lacking. Assays showed from $14 to $75 per ton in gold.

The *Ferris Mine* is situated 6 miles north of Charlotte. There are apparently three veins, the North and South and the Garris. The North vein was worked most extensively. The South vein varies in width from 2 to 7 feet with a pay-streak 18 inches to 4 feet wide. The rich brown ores of the upper levels gave place to the sulphides at depth. The ores carry some chalcopyrite. Assays showed $16.23, $20.41, $28.94, $45.03, $111.62, $128.66 in gold and silver per ton.

Copper also averaged from 13 to 14 per cent. The Garris vein was entered by two shafts, 90 and 120 feet respectively. The vein fissure strikes N. 20° E., and dips 70° N. W. It is reported to vary from 2½ to 5 feet in width. Assays from this vein showed $5.17 and $53.74 in gold.

The *Ray Mine* is 9½ miles southeast of Charlotte. Five veins are reported on this property, the south vein being worked to a depth of 60 feet, and the Phifer Grove vein to a depth of 40 feet. The Ray vein is entered by six shafts, the deepest being 250 feet. The ore seam is 6 to 8 inches thick, and is filled with compact sulphides. It is reported that the ore down to the 150-foot level has been stoped out. Assays showed $20.99, $31.98, $227.11 per ton in gold.

The *Surface Hill Mine* has long been known for its large yield of nuggets. It is situated on a high plateau in Clear Creek township, from which flow McAlpine's Creek to the southwest, Reedy Creek to the northeast, and Clear Creek to the southeast.

The country rock is granite, which is apparently intersected by a system of reticulated quartz veins or veinlets; and these, in consequence of the general disintegration, have scattered their contents widely over and beyond the 66 acres comprising the tract. There appear, moreover, to be two veins of some size crossing each other; viz., the main or Harris vein striking N. 45° E., and the Lidner or Vivian vein striking N. 10° W.

The rich pocket of nuggets, which has given the mine its celebrity, appears to lie near the junction of the two veins, and a little to the north, where a dike has cut across them. It is stated that the nuggets were found most abundantly between the dike and the north end of the Harris vein. It is quite certain that several thousand pennyweights must have come from the space of a few square feet.

A considerable amount of brown ore of good appearance,

carrying copper sulphides, has been mined in the process of hunting for nuggets, but as an ore it is of little value despite its fine appearance. Assays show but $2 to $3 per ton.

CHAPTER VII

THE KINGS MOUNTAIN BELT

GENERAL STATEMENT

The Kings Mountain Belt occupies an area, with imperfectly known boundaries, adjoining the Carolina Igneous Belt on the west.

The country rocks are principally crystalline schists and gneisses with occasional lenticular bodies of siliceous magnesium limestone and beds of quartzite. The strike of the schistosity is northeast, and the dip usually westward at steep angles. There are also occasional dikes of igneous origin. The rocks range in age from Archean to Triassic.

The gneisses are usually micaceous; the schists are micaceous, chloritic, argillaceous, cyanitic, and sometimes graphitic. Pegmatite dikes are frequent in some localities, and near Kings Mountain, Gaston County, and near Lincolnton, Lincoln County, they are tin-bearing. The quartzites are confined to the Kings, Crowders, and Anderson mountains, a line of peaks and ridges stretching from South Carolina in a northeast direction, the highest of which is the Pinnacle, 1,705 feet. The limestone lenses occur along the foothills and lowlands bordering the ridges, in small lenticular masses, usually separated by slates and schists, and often buried without cropping out on the surface.

The gold ores consist of quartz, mineralized zones, and mixtures of limestone and quartz. The gangue minerals are principally quartz, pyrite, with smaller amounts of chalcopyrite and galena. Occasionally lenses of barite are found which carry crystals of galena.

GASTON COUNTY

The *Catawba* (or *Kings Mountain*) *Mine* is located 1½ miles nearly south of Kings Mountain station. This mine is located in a limestone belt, and sometimes the fine-grained schistose limestone was quarried and sent to the stamps.

This mine was discovered in 1834, and worked in a small way for about 40 years. The gold was first discovered in a branch, and in working up the branch to the source of the gold the vein was discovered. It was about 3 feet thick, and contained sugary and cellular quartz with brown ore. Much of the gossan outcrop was too poor to work profitably. At greater depth the ore body was found to be limestone charged with small percentages of sulphides, including a small proportion of galena and the rare lead telluride, altaite. Nearly the whole formation is gold-bearing. The auriferous zone is divided into two parts, separated by a thin seam of talco-chloritic schist, 6 to 24 inches thick, which occasionally dwindles to a mere parting too thin to notice. The entire formation is 60 feet in thickness at places. The front vein has an average width of 11 to 15 feet, which showed assays from 16c to $4 per ton.

The back vein is thought to show more sulphides, but it is not easy to tell the difference between the two when they come together. The limestone is impure, being very siliceous and highly magnesium. The big chimney of ore, which gave the mine its reputation, had a well defined pitch of its own in the vein, northward. It is reported that the great width of the ore bodies, the ease with which the ore was mined and milled, and the small amount of sulphides combined to make the treatment of the ore, even though low grade, profitable. One time 40 stamps were in operation, and it is reported that during its history the mine produced $750,000 in gold. Assays on the ore showed $12, $5.13, $3.15, $38.54, and $74.43 per ton in gold.

The *Crowder's Mountain* (or *Caledonia*) *Mine* is located 4 miles east of the Catawba Mine, and on the east side of Kings Mountain. The country rocks are sericitic and chloritic schists, sometimes silicified, and often ferruginous.

Certain narrow zones or belts of the schists are slightly mineralized with iron and copper pyrites. Occasionally the width of the ore-bearing zone reaches a width of 8 to 10 feet. The ores are commonly low in grade. The results from a number of assays are as follows: $5.17, $1.03, $3.62, $9.10, $9.10.

The *Rhodes Mine* is 18 miles southwest from Charlotte. The ore body is auriferous mica gneiss, and has been worked to a depth of 100 feet, and for a length of 300 feet. Galena, the lead sulphide, is sometimes found.

The *McLean (Rumfeldt) Mine* is 15 to 16 miles southwest from Charlotte. It has been prospected probably for a length of 200 yards, and to a depth of 110 feet. The vein is filled with quartz carrying iron pyrites, and is from 1 to 6 feet in width. Some placer ground is found on the property.

The *Duffie Mine* is 16 miles southwest of Charlotte on the old Tuckaseegee road. The vein is from 2 to 10 feet wide, and has been worked to a depth of 110 feet, at which point a large body of low-grade sulphides was found. Some assays show the following values: $4.14, $5.17, $4.92, $10.33, $12.40, $13.43, $16.54 per ton with a small trace of silver.

The *Oliver Mine* is situated 12 miles southwest from Charlotte, on the west side of the Catawba River. It is believed to have been among the earliest operated mines in this section, and there are traditions of work here which was done prior to the Revolutionary War. This mine has been worked for a distance of 100 yards, sulphides appearing at a depth of 75 feet, especially galena, which was rich in gold. No assays are given showing the values at this mine.

The *Long Creek Mine* is situated 6 miles northwest of Dallas. There is said to be 3 veins, the Asbury, Dixon, and McCarter Hill. The country rocks are principally schists, which strike N. 20° to 25° E., and the dip generally 85° N. W. The strike of the vein is conformable to the schistosity.

The *Asbury* vein has had some extraordinarily rich shoots of ore, which carried iron and copper pyrites, galena, sphalerite, mispickel, and carbonate of bismuth. Its width was from 6 to 8 feet. It has been opened by two shafts, 45 feet apart, and worked to a depth of nearly 140 feet.

The *Dixon* vein has been worked extensively along the surface by pits and two shallow shafts, 300 feet apart, from which drifts were run 79 feet south and 107 feet north. The thickness of the vein was a little over 3 feet.

The *McCarter Hill* vein has been entered by 3 shafts within a distance of about 250 feet and was stoped to a depth of 160 feet in the ore shoot, which pinches to the north, and increased longitudinally until at the 140-foot level it had a length in the vein of more than 211 feet; the width ranges from 4 to 6 feet.

The last work done at this mine was in 1892, but it is reported that some recent work was done by the American Smelting & Refining Company. The results of this last work are not known. Previous work showed values of approximately $8 per ton.

LINCOLN COUNTY

No mining for gold has been done in Lincoln County for several years other than some recent exploration work in the southern part of the county.

The *Hope Mine,* near Lincolnton, has been developed to a depth of 110 feet with drifts which have been run for some length.

The *Graham Mine* is situated about 4 miles northeast of Iron Station. The vein is reported to be 30 to 42 inches wide, and has been prospected by pits for about 100 feet on the outcrop. The ore usually contains copper and occasionally is classed as copper ore.

CATAWBA COUNTY

Only one mine has been developed to any extent in Catawba County, even though considerable prospecting has been done from time to time. Some prospecting was done in 1934, but nothing of value was found.

The *Shuford Mine* is located 4½ miles slightly south of east from Catawba on the Southern Railway. The underlying schists and gneisses are cut by numerous seams of gold-bearing quartz, which run more or less in all directions. The surface of the ground is covered with auriferous quartz, and in some instances the soil is also auriferous. The entire surface in the beginning was considered as "pay" material. The mine is best adapted to a combination of hydraulicking and milling methods.

The last work accomplished at the Shuford Mine was the explorations of the quartz seams at depth, some of which reached 12 inches or so in thickness, but none were found to contain such value as to be worked profitably.

At a great number of localities within the immediate vicinity of Maiden and Newton some prospecting has been done, but nothing has been developed worth while.

DAVIE COUNTY

The rocks underlying Davie County are principally gneisses, which strike more or less in a northeast-southwest direction. At a number of localities gold has been mined to some extent, but with what results is unknown. The most important localities are the *Butler* or *County Line Mine,* the *Isaac Allen Mine,* and several places on Callahan Mountain. Most of the deposits worked are reported to be of low value.

YADKIN COUNTY

The *Dixon Mine,* discovered in 1894, is 8 miles southeast of Yadkinville. The country rock is principally a decom-

posed mica schist, sometimes chloritic, occasionally intersected by diabase dikes.   The vein is quartz, and the outcrop
shows a width of 4 feet.   It dips steeply to the northwest.

The shaft was sunk to a depth of 35 feet, at which point
the width of the vein was 4 feet.   At this level drifts were
sent both ways for a length of 50 to 60 feet.   At the end of
the drifts the vein runs off into the laminations of the
schists, making lenticular deposits and stringers.   The value
of the ore is not uniform, as assays show values from $7 to
$40 per ton.   The mill results, however, gave an average of
approximately $5 per ton.

There are no other important occurrences of gold in
Yadkin County, even though in 1934 news articles stated
that some prospecting and development work was accomplished near Elkin.   Apparently nothing of value was discovered, as the exploration work was soon abandoned.

CHAPTER VIII

The South Mountain Belt

GENERAL STATEMENT

The South Mountain range forms one of the most prominent eastern outliers of the Blue Ridge, in Burke, McDowell, and Rutherford counties. This range is the nucleus of one of the important gold-bearing belts in the State, and comprises an area of 250 to 300 square miles. It extends from Morganton on the north to Rutherfordton on the south, a total length of about 25 miles, with an average width of 10 to 12 miles. Included in this same belt, however, are those isolated deposits which occur north and east of Morganton, in Caldwell and Wilkes counties, as well as those to the south of Rutherfordton in Polk County.

TOPOGRAPHY

The South Mountain range extends in a northeast-southwest direction, with a maximum elevation of 3,000 feet above sea level, and an average elevation of about 1,300 feet.

The main drainage is in a northern direction, and all streams empty into the Catawba River at an elevation of approximately 1,100 feet. The principal streams flowing northward are Silver and Muddy creeks, with their tributaries dissecting the old plateau. The principal stream, rising in the South Mountain range, which flows southward is the First Broad River. The headwaters of this stream rise in the Golden Valley section of the South Mountain range.

The above named streams supply water for hydraulicking purposes.

GEOLOGY

The principal rock formations of the South Mountain region are mica and hornblende gneisses and schists. These rocks have an eminently lenticular structure, which strikes

N. 10° to 25° W., and dip 20° to 35° N. E.   The general strike and dip, however, apply only to the central part of the region.

The gneisses and schists contain, in addition to the mica and hornblende, especially in certain localities, a high garnetiferous content, with many rare minerals, such as zircon, monazite, xenotime, and fergusonite. These gneisses and schists are apparently altered igneous rocks, especially granitic and dioritic, of Archean age.   These rocks have been rendered schistose by the action of various dynamic forces.

All of the rocks in this belt have been weathered to considerable depth, and only occasionally are good outcrops observed.   The decomposed layer has an average thickness of probably 50 feet or more, with occasional thicknesses of 100 feet or more.   This subaerial decay of the rocks is almost universal and causes great difficulty in studying the lithology of the section.

The gneisses are composed essentially of quartz, feldspar, and mica.   Both muscovite and biotite are present, the latter usually predominating.   All of the feldspars are present at various times.   Hornblende and pyroxene are altogether absent, or present only to small extent, in the acid mica gneisses.

Alteration products are numerous.   The micas are frequently altogether gone over into hydrous varieties and carbonates; and bleached biotites are common.   The feldspars are always affected, partially or completely altered to kaolin or sericite.   The ferro-magnesium minerals alter to chlorite and epidote.

Due to the alterations, the original character of the gneisses and schists is entirely obliterated. The complete obliteration of the original structure makes it almost impossible to trace any particular formation a great distance.

The dikes cutting the formations are both acid and basic.

The acid dikes are principally granite and pegmatites, while the basic are principally diorites and gabbros. The true diabase, as found in the Piedmont section, is absent in the South Mountain region. However, in the north extension of the belt an olivine-diabase dike is found. Isolated masses or lenses of pyroxenite and amphibolite are occasionally found. These are usually coarse crystalline and massive, devoid of schistosity. These dikes apparently were intruded after the force producing schistosity was exerted. These masses are looked upon as basic segregations from the original igneous magma out of which the gneisses were formed.

QUARTZ VEINS

The quartz veins of the South Mountain region are true fissure veins. The fissure system is the most regular, persistent, and remarkable, from a point of almost absolute parallelism, in the State. The strike is N. 60° to 70° E., and the dip is from 70° to 80° N. W.

The width of the veins varies from very thin to 4 feet or more. The smaller veins, however, contain the higher values, and veins are seldom found which are thick enough to pay to work. Occasionally, as has been the experience of miners, the quartz veins are sufficient in number and contain such values to pay to hydraulic the entire decomposed surface of the weathered portion. Sections of this type have been completely worked out, and it is not likely that any will be discovered in the future which can be worked profitably, except at the heads of valleys where water is not available.

The quartz of the fissure veins is usually milky white in color, generally saccharoidal (sugary), and seldom, if ever, vitreous or glassy. The quartz is often stained brown and is cellular from the decomposed sulphides. It is universally true that below water level sulphides are encountered, especially pyrite, galena, chalcopyrite, and sphalerite. Apparently the vein matter was formed by ascending mineralized solutions.

There is no evidence of replacement of the country rock by ore.

It is impossible to give with any degree of accuracy the true value of the quartz veins found in the South Mountain region. Occasionally samples are found which show values running several hundred dollars per ton, but this is unusual. A pocket of ore was found at the Hercules Mine in a 30-inch quartz vein, which showed enormous values. It is reported that one flour sack of the high-grade ore brought $1,500. Such pockets, however, are seldom found.

The average value of the quartz veins, especially the smaller veins, is from $5 to $20 per ton. The placer deposits in this region are formed by the accumulation of the quartz from the breaking down of these veins.

The fineness of the gold from this section varies from 825 to 950, distributed as follows: Brindletown district, from 825 to 850; Vein Mountain, from 780 to 800; Golden Valley, 900; and in Polk, 900 to 950.

THE QUARTZ VEINS OF THE SOUTH MOUNTAIN REGION

Generally speaking, quartz veins in the South Mountain region are too small to be worked profitably except at a few localities. To date no quartz vein has been worked on a large scale in this section, and there is doubt as to the discovery of any veins in the future which would justify large scale operations.

At a great number of localities the natives, both men and women, hunt for the small quartz veins or "rich streaks" of the small veins, from which they extract the gold. It is stated that frequently the natives average from 50c to $1.50 per day, after paying the owners a toll or royalty of one-sixth.

The rich quartz is crushed in a mortar, the crushed material panned carefully, and the gold collected with mercury.

The vein mining in the South Mountain region will probably continue to be desultory and spasmodic. The nature of the veins so far discovered does not offer profitable operation on a large scale.

BURKE COUNTY

The *Brown Mountain Mine* is situated 13 miles north of Morganton on Kingy Branch, a tributary of Upper Creek. The main mass of Brown Mountain is made of granite, while its westward slope, towards Upper Creek, is composed of chloritic schists. The work done at the property consisted chiefly of prospecting; No. 1 shaft was sunk only 20 feet deep, and the crosscut was only 5 feet deep.

The quartz veins appear to lie in irregular stringers in the granite, varying from 1 to 6 inches in thickness with only one vein reaching 2 feet in width. Assays from the quartz show from a trace to $12 per ton.

CALDWELL COUNTY

The *Hercules Mine* is located 12 miles north of Morganton. In 1900 some development work was carried on by Mr. Robert Ore, of Newport News, Virginia. In the beginning very favorable and promising results were obtained for the amount of work done.

A shaft was sunk to a depth of 50 feet on a quartz vein, which was 8 inches wide at the top but widened to 30 inches at the bottom of the shaft. This vein was traced on the surface for nearly ½ mile by means of open cuts and pits. At the 25-foot level a drift of 100 feet was run on the vein from the shaft, and it was found to be continuous throughout this distance. The ore that was taken out showed up remarkably well.

In 1930 the shaft was sunk to a depth of 95 feet, at which place the vein was from 30 to 36 inches in width, and a very rich pocket of high-grade ore was encountered. The property was reported sold, but little or no development work was carried on. The property is idle at the present time.

The *Baker Mine* is situated on the western slope of Davis Mountain, near the river.

The country schists and gneisses strike northeast, and dip southeastward. The quartz veins are in the schists and run northeast and southwest. A diabase dike apparently crosses the vein, and it is not known whether the vein continues into and across the dike. Sufficient development work has not been done to determine this fact. There are four principal veins; the more northerly is the Braswell, which has a large projecting outcrop, but has never been examined. The Goley Ann vein is narrow at the surface. The shaft vein, to the southwest, is from 20 to 24 inches in width, and has been opened at points 200 yards apart. The quartz carries scattered galena and associated minerals.

The *Cabin vein* is still further south, and was worked at points 100 yards apart, and is said to be 20 inches in width. The vein is quartz with ferruginous matter and galena in considerable quantities. The galena is auriferous as well as argentiferous. The quartz of all the veins, aside from the galena, carries gold.

Assays of the pure material showed values of $108.55, $121.89, in gold and silver. The lead content was approximately 84 per cent.

The *Bee Mountain Mine* is situated 4 miles northwest from Lenoir on the northeast slope of Bee Mountain. The country rock is a garnetiferous mica gneiss with pegmatite intrusions. The openings consist of two shallow prospect shafts and a tunnel. No. 1 shaft is 70 feet deep, and No. 2 is 30 feet deep. The quartz vein is said to be 4 feet in thickness, and much cellular and brown stained quartz is found. In addition to the gold, some lead, zinc, and copper is also found. A tunnel was sent into the hillside for a distance of 100 feet, but it did not intersect the vein.

The *Miller* and *Scott Hill mines* adjoin each other, and are situated on the waters of Seley's Creek, 1½ miles northwest of Hartland.

The float quartz and that obtained from the old diggings is a white vitreous variety, and the largest pieces are 12 inches in thickness. Little is known of the results obtained in the cuts, tunnels, and shallow shafts.

MCDOWELL COUNTY

The *Marion Bullion Company Mine* is located at Bracket-town in the valley of the headwaters of South Muddy Creek.

The country rock is biotite gneiss striking N. 10° W., and dipping 10° to 15° N. E. It is lenticular in structure, and incloses lenses of high altered greenish sericite gneiss.

A number of small quartz veins were prospected on the property. The most extensive explorations in this direction were made several years ago by the sinking of a vertical shaft, 7x11 feet cross section, to a depth of 126 feet on a series of six narrow quartz veins lying close together.

Near the outcrop the quartz veins vary from 1 to 6 inches in thickness, and are from 1 to 3 feet apart. They are more or less parallel, and strike N. 6° to 8° E., and dip from 63° to 70° N. W. The primary object of the shaft was to ascertain whether these veins would grow large and come together in depth, according to a popular and fallacious belief; instead of this, they are found to pinch out to less than one inch at the bottom of the shaft and to maintain their distance apart. Two small normal faults were observed in the veins, of as much as ½-inch throw. The quartz is saccharoidal and mineralized with galena, sphalerite, chalcopyrite, and some pyrite. The wall rock itself, though not highly so, is impregnated with pyrite to a greater extent than the vein matter.

The quartz alone from these veins gives assay values ranging from $4 to $20 per ton in gold and silver. At the depth of 115 feet a sample of the entire shaft material, vein quartz and wall rock, assayed $4.80. A test mill run was made in September, 1894, in a Huntington mill and two Frue vanners, situated on the property. Fifty-four tons of quartz and rock were treated; of this amount one-fourth was quartz from the 50-foot level in the shaft, three tons were float quartz, collected in hydraulicking on another part of the property, and the remainder was rock and vein quartz from the shaft. The result was 71 dwts. of free gold, caught as amalgam in the body of the mill beneath the stationary grinding plate, and on the Hungarian riffles; practically no gold was obtained from the silvered copper plates. The concentrates from the two Frue vanners amounted to 2 tons, which were so dirty that they were run over and reduced to 260 pounds. The tailings on this second run showed enormous loss, especially in galena and floured amalgam, due either to the inefficiency of the vanner or more probably to inexperience in operating the same. An assay of the final concentrates showed only $7.88 per ton, and a sample of the tailings gave $2.63. This is certainly discouraging. If further and more extended exploration can show that the entire rock mass, at least over a considerable width, in the bottom of the shaft and in the cross cuts, will average from $3.50 to $4.50 per ton, then by careful and intelligent management the mine might be a profitable low-grade proposition; but certainly no dependence can ever be placed on the small veins alone.

The *Vein Mountain Mine* is located on Second Broad River between Vein Mountain and Huntsville Mountain.

The crystalline schists at Vein Mountain have a general strike of N. 10° to 20° W., and dip of 30° N. E. A number of small granite dikes cut the country rock at various angles, with unmistakable signs of faulting.

A series of as many as thirty-three parallel auriferous quartz veins crosses at Vein Mountain in a belt not over ¼-mile wide. The principal and largest one of these is the "Nichols" vein, which has been prospected in four shafts, the deepest one 117 feet. Shaft No. 1 is 1,200 feet east of No. 4, and 100 feet above it in elevation. The strike of the vein is about N. 80° E., and its dip varies from 75° N. W. to nearly vertical. Its thickness is reported to vary from a few inches to 3 feet, the usual portion of the vein worked being from 15 to 30 inches. Below the water level the quartz is mineralized with sulphides, chiefly pyrite and chalcopyrite. A number of assays show the following variation in the value of the quartz:

Assays, Gold Ores, Nichols Vein, Vein Mountain.

Gold, per ton	$2.58	$4.17	$6.20	$10.33	$13.43	$12.40	$70.28
Silver, per ton	Trace	Trace	Trace	Trace	Trace	1.51	5.43
	$2.58	$4.17	$6.20	$10.33	$13.43	$13.91	$75.71

A former superintendent, Mr. B. G. Gaden, stated that the ore averaged between $15 and $17 per ton. A 10-stamp mill operated for some time, but it is impossible to state why the mill closed, unless it was impossible to recover the gold from the partially oxidized sulphides by straight amalgamation. There was no concentration equipment on the property in connection with the stamp mill.

RUTHERFORD COUNTY

The *Alta* (*Monarch* or *Idler*) *Mine* is situated about 5 miles north of Rutherfordton, on the divide between Cathey's Creek and the Broad River. As many as thirteen parallel quartz veins have been explored on this property, within a distance of ½-mile across the strike. The four larger veins encountered are known as the Monarch, Alta, Carson, and Glendale. These various veins were worked from 1845 to

1893 or 1894 by numerous shallow open cuts, pits, and shafts, down to water level; the ore was milled in arrastras. Apparently the last work was done on the Alta vein about 1894. A shaft was sunk on this vein to a depth of 105 feet, but little is known as to the nature of the vein. It is reported that the vein strikes N. 65° E., and the thickness varies from 10 to 22 inches, and averages perhaps 15 inches. The quartz is of a milky variety mineralized with pyrite and some chalcopyrite. The ore contained from 1 to 20 per cent sulphides, and averaged about 5 per cent. One statement gave its average yield in free gold, by mill tests, at $10 per ton; another gave $30; and the value of the concentrates is said to have averaged $100 per ton. A 5-stamp mill operated at the property for some time.

The *Ellwood Mine* is situated 3 miles N. 20° E. from Rutherfordton and 1½ miles southwest from the Alta Mine, on the waters of Cathey Creek. A series of five parallel quartz veins, 100 feet and more apart, was first opened in 1842. The last work was done in 1893, but the owner stated that he barely paid expenses. However, the ore was reported to run from $5 to $7 per ton in free gold, while the sulphide ore was reported to run $20 per ton. The thickness of the larger veins varies from 10 to 15 inches.

The *Leeds Mine* is situated on the quartz vein parallel to and 100 feet north of the Ellwood veins. This mine was abandoned previous to 1892 and has long been inaccessible; consequently little is known of the quantity and quality of ore.

POLK COUNTY

The gold mines in Polk County are extensions of the deposits in the South Mountain area. The principal work accomplished in this county was on placer material, but two or three quartz veins were opened up to some extent.

The *Double Branch Mine* shows a small quartz vein, 1

to 3 inches in width. The assays varied considerably from $2.07 to $465.07 per ton.

The *Red Spring Mine* is reported to have three small veins, but little is known about them.

The *Splawn Mine* is reported to have a massive vein of low-grade quartz.

The *Smith Mine*, ½-mile east of the above, consists of a series of narrow quartz veins quite similar to those at the Carolina Queen Mine.

WILKES COUNTY

The *Flint Knob Mine* is located 6 miles east of Deep Gap. The ore is principally a silver-lead with traces of gold. This property was explored to some extent between 1830 and 1840. Sufficient work has not been accomplished to determine the nature of the vein.

Near Trap Hill, on the eastern face of the Blue Ridge at Bryan's Gap, there is a bold outcrop of quartz, which has been traced for nearly 3 miles. The vein varies from 3 to 20 feet in width and carries pyrite with a small portion of chalcopyrite. The sulphides carry small amounts of gold. Assays showed $1.57, $2.07, $3.10, $4.14, $5.17, $10.29 per ton, with a trace of silver.

CHAPTER IX

THE GOLD DEPOSITS WEST OF THE BLUE RIDGE

GENERAL STATEMENT

At a number of localities west of the Blue Ridge Mountains there are several gold mining localities, which have attracted, in the past, some attention. Most of these deposits, however, have shown little production even though many of them have been prospected and explored to considerable extent. Among the more notable localities are those on the southeastern slope of Forge Mountain in Henderson County, the gravel deposits in Fairfield Valley, and the gravel deposits in Valley River Valley in Cherokee County. In the counties of Ashe, Alleghany, Jackson, and Swain the producing copper mines have shown low gold and silver values. The highest values in gold obtained from the copper ores are from the copper ores of the Copper Knob Mine in Ashe County. It is generally true that the sulphides occurring in the quartz veins and schist zones in the Mountain Region carry gold values. Especially is this true of the veins carrying pyrrhotite in Swain County in the vicinity of Bushnell. Even though many of these sulphide deposits carry gold, generally speaking they are not regarded as possible producers of precious metals.

ASHE COUNTY

The *Copper Knob* (or *Gap Creek Mine*) is located in the southern part of Ashe County on the waters of the New River. At this mine there are three quartz veins, only one of which has been worked to any extent. This vein lies in a large body of hornblende schist, but the prevalent country rock is a gray gneiss, with a strike N. 60° E. and a dip 40° S. E.

The vein is principally quartz, which, in the upper part, contained a great deal of iron oxide, extraordinarily rich in gold. The mineralized seam occupied the center of the quartz fissure and varied from 4 to 6 inches in thickness. At the 60-foot level, the vein carried copper ores for 53 feet; then the quartz changed slightly and carried native gold, with brown iron oxide, for a distance of about 36 feet. At this point the strike changed to N. 43° W. The vein seemed to be more mineralized on the northwest than on the southeast side of the workings. The ore is complex: vitreous copper ore, malachite, chrysocolla, a very little chalcopyrite, and brown ore. The iron pyrite is almost wanting. The ore seam increased somewhat in width as a greater depth was reached. The shaft was sunk to a depth of 60 feet with satisfactory results, and subsequently deepened to 140 feet. At this stage the mine became the prey of a company of speculators. Those persons familiar with the last work stated that the mine continued good as far as explorations extended.

The nature and value of the ore is shown by the following assays: $8.62, $34.79, $57.36, $77.51 in gold, and $2.26, $25.50, $14.53, $45.68 per ton in silver. The copper content varied from 23 to 37 per cent.

WATAUGA COUNTY

In Watauga County, on Howards Creek, some gold gravel has been found. At this locality some work was done on a small scale before the Civil War.

One mile north of Boone, on Hardings Creek, some gold has been found.

BUNCOMBE COUNTY

Some indications of gold have been found on Cane Creek.

In 1935, on the property of Dr. P. B. Ore of Asheville, some quartz veins were explored to some extent, but the veins were narrow and the gold content low. The work was soon abandoned.

The *Boylston Mine* is situated on the southeastern slope of Forge Mountain, on Boylston Creek, 12 miles west of Hendersonville.

Forge Mountain is made of fine-grained mica and hornblende gneisses and schists, in part much crumpled. The general strike of the schistosity is N. 20° to 30° E., and the dip is to the northwest.

The schists are cut by granite dikes, the general strike of which is about N. 30° E., dipping apparently at a steep angle. It is a light-colored granite, coarse-grained, containing much biotite, and near the surface is largely decomposed to white-gray. The width of the dike is not definitely known, but appears to be over 100 feet.

The valley of Boylston Creek is composed of schistose limestone, which strikes N. 40° E., and dips 45° S. E. The limestone is usually overlain with schist in which are found rusty quartz lenses from $\frac{1}{2}$ to 2 inches in thickness, which carry traces of gold and silver.

Four main quartz veins have been discovered on the property. These are more or less parallel, and strike N. 30° E. In places the quartz is deeply corrugated, and has a banded structure. The character of the quartz is usually crypto-crystalline, vitreous, being cellular and stained brown above the water level. It is rarely fine-grained, saccharoidal, and milky.

The thickness of the veins varies from 1 to $4\frac{1}{2}$ feet, and the paystreak is usually on the hanging wall and is from 1 to 8 inches wide. Near the surface the quartz carries free gold, but below water level it contains sulphides, pyrite and galena.

The No. 1 vein is reported to be $2\frac{1}{2}$ feet in width, with $2\frac{1}{2}$-inch seam of slaty material. Assays on this vein showed $4.13, $13.37, $13.44 per ton in gold, with little silver.

The No. 2 vein has been more prospected and developed than any of the others. All of the ore which has been milled on the property was obtained from this vein. Openings have been made at intervals on this vein over a length of 1,500 feet. The vein is exposed in open cuts, and shows a width 3 to 4 feet thick composed of sugary quartz and decomposed sulphides. A rich paystreak, from 1 to 3 inches thick, of reddish-brown quartz is found on the hanging wall.

It is estimated that 1,000 tons have been taken from this vein, which, with an additional 2,000 tons above water level, was reported to contain $4 per ton in gold. A number of assays from this vein showed the following values in gold per ton: $7.23, $4.13, $4.13, $5.18, $6.20, $7.30, $109.25, $84.75, $7.23. The two high values shown were samples taken from the rich paystreak. In addition to the gold, the ore ran from 41c to $4 per ton in silver.

A number of openings have been made on the other veins on the property with the following results: $4.13, $4.13, $4.13, $23.77, $2.06, $9.09, $8.77, $19.10, $4.63, $5.58, $5.59, $6.70, $6.52, $11.37, $18.60, $3.10 in gold per ton, with the silver showing from 32c to $2 per ton. Assays on the rich paystreak showed values from $34 to $55 per ton.

The property was at one time equipped with a stamp mill which had a capacity of 10 tons per day but saved only 24.63 per cent of the assay value. The low yield was due to the condition of the machinery and the battery plates.

During the last five-year period two or three attempts have been made to develop the deposits, but without success. In the summer of 1935, under the direction of Marshall Cravatt, Mining Engineer, of Asheville, North Carolina, further exploration work was begun. Recent news reports stated that favorable results were obtained during the prospecting and development work and that plans were under way for continuous operation.

TRANSYLVANIA COUNTY

The most important gold-bearing locality in Transylvania County is in Fairfield Valley on Georgetown Creek. Reports state that between $200,000 and $300,000 in gold were recovered from the gravel along the stream. The reports also state that the deposits extend for several miles and have been by no means exhausted.

The origin of the gold in this valley is from the quartz veins cutting the gray gneiss which rises on the northeast as a precipitous wall to a height of 700 or 800 feet above the valley. It is along the base of this wall, where Georgetown Creek has cut a deep channel, that the gold has been principally obtained.

JACKSON COUNTY

There has been a small gold production from Jackson County from placer or detrital beds. These localities are situated chiefly along the southern slopes of the Blue Ridge, near Hogback, Chimney Top Mountains, and in Cashier's.

CHEROKEE COUNTY

In Cherokee County, the gold belt is in the same body of soft slates and schists which carries the limestone and iron. It is found in both vein and superficial deposits. The sands of Valley River yielded profitably through a large part of its course, and some very rich "washings" have been found along its tributary streams on the north side. The origin of this gold is very near the limestone. A remarkably rich vein was opened near the town of Murphy, known as No. 6, which immediately underlies the marble. The No. 6 vein was composed of quartz carrying silver and lead, in which was imbedded a large per cent of free gold.

Southeast of the limestone is a series of "diggings" along the lower slopes of the mountains from near Valleytown to Vengeance Creek, a distance of 12 to 15 miles. The gold is found in the drift, which covers the lower spur and terminal ridges of the mountains lying south of Valley River. The drift beds have a depth of from 10 to 20 feet, and an elevation above the river of 150 to 200 feet; they are remarkable for the great size of the quartz boulders and the very large and abundant staurolite, "fairy cross" crystals.

During the past five years considerable exploration work was conducted in the Valley River section of Cherokee County, with little or no success. A group from Murphy spent between $3,000 and $4,000 in exploration work but soon abandoned the property. In the summer of 1935, some further development work was undertaken under the direction of mining engineers, but the results of the venture are not known.

CLAY COUNTY

The belt of gold deposits which occurs in Cherokee County extends beyond the Hiwassee into the Brasstown Creek section of Clay County.

During the past year or two M. R. Hilford, Hendersonville, conducted some exploration and development work at the Warren Mine on Brasstown Creek. The quartz vein, standing almost vertically, was exposed in a shaft about 40 feet in depth. Apparently the quantity and quality of the ore encountered showed considerable promise, as a small 10-stamp mill was shipped to the property. However, due to some misunderstanding, the stamp mill was never erected. Little or no work has been accomplished since that time.

CHAPTER X

GOLD PLACER DEPOSITS IN NORTH CAROLINA

GENERAL STATEMENT

The gold placer deposits in North Carolina may be divided into four classes, as previously outlined, depending on the nature of the material. The four classes are: (1) The gravel beds along the stream and adjoining bottom lands, deposited by stream action; (2) bench gravels on the hillsides, deposited by stream action but left well above the present stream levels due to erosion; (3) gulch and hillside deposits due to the secular disintegration and the resulting accumulations induced by gravity and frost action, which may be classed as residual gravels; (4) the talus accumulated at the base of prominent hills or slopes and other prominent topographic features; (5) the upper decomposed layer of country rock, with the rotten rock in place, known as saprolite. This last group of saprolite deposits is classed under placers, due to the methods of working such deposits, which are similar to placer methods.

STREAM GRAVELS

Along many of the streams in North Carolina there are many small deposits of stream or river bottom gravels which contain gold. These gravel deposits are largely confined to the valley flats on either side of the streams. The gravel beds seldom exceed 3 to 4 feet in thickness, and usually have an overburden from 2 to 10 feet, composed essentially of clay, which is barren. The gravel is more or less rounded and usually rests upon the upturned edges of the country rock. Occasionally the overburden may contain a small amount of fine gold, but it is seldom in sufficient quantity to pay to work. In some sections, where basic dikes intersect the country rock, fragments of these dikes are included in the stream gravels.

BENCH GRAVELS

In many sections in North Carolina, on either side of the streams and well up on the hillside, there are small deposits of gravels, which may be considered as bench gravels. These gravels were originally deposited along the valley bottoms by old streams, which later, during the long process of erosion and the subsequent deepening of the valleys and the meandering of the streams, were left above the present valley flats. This is especially true in the South Mountains region, where these gravels have produced considerable amounts of coarse gold. These deposits were worked over by old-time prospectors and miners time and time again, and it is doubtful that any deposits remain which are of such extent and quality to be commercial.

The gravel in these bench deposits consists largely of quartz with large boulders of the dike materials. The gravel is more or less rounded due to stream action. These deposits usually contain considerable overburden, as fine sand, clay and soil, quite similar to the material covering the stream gravels.

RESIDUAL GRAVELS

In the lower Piedmont section of the State, especially in the sections where the topographic relief is approaching peneplaination, there are deposits of angular gravels, which may be classed as residual gravels. These deposits are confined largely to small gulches and valley heads. These gravels are angular and are deposited by gravity and frost action. The quartz composing these gravels are fragments from the weathering of the quartz veins in the immediate vicinity. In some sections of the State, as the Portis and Parker mines, these residual gravels, accumulated in the shallow valleys, have been worked rather extensively with good results. The deposits have been worked over and over again, but by no means should these old workings be overlooked, as some of them have possibilities.

The residual gravels are more or less mixed with fine sand and clay, and as a result there is some concentration of the gold on the surface with somewhat greater concentration on the bed rock. In sections where the clay content is high the gravels have been untouched because to date no successful method of recovery has been worked out.

TALUS DEPOSITS

In the upper Piedmont or more rugged sections of the State, there are gold deposits, which may be classed as talus deposits. These deposits are found at the foot of the high hills or mountains on the Pilot Mountain section in the South Mountains region and the Fairfield section in Transylvania County. The deposits occur on the hillsides and more gentle slopes, and are formed by the accumulation of the fragments of the small quartz veins. They usually contain fragments of the country rock as well as clay and soil. They are never horizontal in position but are at some angle, usually approaching the angle of repose. Many of the deposits have considerable overburden, consisting especially of clay and soil, due to hillside creep and frost action.

In certain sections these talus deposits have been worked by underground methods, as it was too expensive to remove the overburden. In mining deposits of this type considerable timber is necessary to prevent the caving of the overhanging materials. In the Pilot Mountain section of the South Mountains region there are numerous old tunnels, which have penetrated these deposits to a depth of close to a hundred feet. Many of these tunnels are still in existence, and it is surprising how some of them have stood up for many years without being timbered.

Apparently, from the nature of the old workings, the gravel was hauled by wheelbarrow to the nearest streams and there washed. Large piles of gravel have been screened out and stacked as if the miners had planned to grind it and recover the gold. These gravels have shown up to $5 per ton in gold, but it is doubtful if any of the gravels contain such an average to pay to mill. Some of the gravels have been worked as many as five times, which is probably due to the fact that the early methods of recovery did not recover all the gold.

SAPROLITE DEPOSITS

In the Piedmont section especially, there are gold deposits which may be classed as saprolite deposits. Saprolite is the decomposed rock in place, which is more or less mineralized. The upper surface of the bed rock has been completely decomposed, but no movement has taken place, as the weathered portion shows the structure of the rock. The gold in the saprolite deposits is usually free milling and seldom averages more than a dollar or two per ton. The gold distributed in the saprolite may or may not be due to fine quartz stringers which cut the decomposed rock in various directions. Sometimes, however, the stringer veins conform to the schistosity of the country rock.

The gold content of the saprolite usually averages higher near the surface due to the accumulation of the gold from the weathering of the vein material. The gold concentration on the surface is due to the lighter materials being removed by surface waters. The decomposed rock also contains sufficient clay to prevent penetration by the gold particles. The gold values are sometimes confined to regular zones and are not distributed uniformly over wide belts.

DISTRIBUTION OF PLACER DEPOSITS

The most important stream placer deposits are those of Silver and South Muddy creeks and their numerous tributaries, in Burke and McDowell counties; the First Broad River and its tributaries, in Rutherford County; and the Second Broad River with its tributaries, in McDowell and Rutherford counties. These streams have their source in the South Mountains region. Since the source of the gold is from the numerous quartz veins near the headwaters of these streams, it has been transported considerable distance by the streams.

The most accessible stream deposits, especially those where water is available, have been exhausted. The placer mining in the past was confined principally to the deep gravel channels, the gulches, and, to some extent, the decomposed country rock or saprolite. The distribution is more or less general along the streams, and the principal centers of operation were at Brindletown, Bracket-town, and Vein Mountain.

Another section which has possibilities is that section underlain with gravels, along Shocco and Fishing creeks and their tributaries, in Nash, Franklin, Halifax, and Warren counties. In this section the gravels are confined principally to the valley flats. On the south side of the streams there are also deposits which may be classed as saprolites, which offer possibilities.

In Randolph and Montgomery counties, especially on Uharie and Little rivers and their tributaries, there are deposits of gravel as well as saprolites which have possibilities. In other sections of Montgomery and Randolph counties, as well as the northwestern part of Moore County, there are numerous saprolite deposits which are worthy of further investigation.

In addition to the above sections, there are isolated saprolite deposits in Stanly, Union, and Catawba counties, which show low values but may be worked profitably if properly handled.

VALUE OF PLACER DEPOSITS

It is impossible to give any definite information as to the value of the placer deposits in North Carolina. These deposits are usually spotty, of indefinite value and quantity, and it is only through careful investigations that the values can be determined. Records show that they are quite variable, ranging from a few cents to as high as $20 per cubic yard. Generally speaking, however, the values are less than $1 per yard. In the vicinity of the Portis Mine, in Nash County, some of the gravel deposits show from 10c to 50c per cubic yard, while the saprolites show from $1.50 to $12 per yard, with an average of $2 to $3. Numerous assays have been made by various companies on the Portis property, which show an average better than $2 per yard.

At the Parker Mine, in Stanly County, the placer and saprolite deposits show a gold content from 10c to $2.50 per yard. However, the values are not uniformly distributed, as there seems to be a concentration on the surface from 4 to 6 inches in depth with a further concentration on the bed rock, with little or no value between.

Saprolite deposits in Montgomery and Randolph counties show values of 50c to $3 per yard, usually rather spotty, and it is impossible to give the average of the deposits until further prospecting is done. In some sections, however, there are possibilities for commercial production.

In the placer and saprolite deposits the gold is usually fine; although in certain localities, as the Reed, Parker, and Portis mines, some coarse gold is found. Some very fine nuggets have been found at the Reed and Parker mines.

METHODS OF WORKING THE PLACER DEPOSITS

Due to the nature and distribution of the placer deposits in North Carolina, every method known has been used in an attempt to recover the gold from the placer materials. These methods included the hand panning, sluice boxes, rockers, hydraulicking, log washers, Snodgrass machines, trommels, centrifugal machines, and, in three instances, dredges. A great many of the processes used have been failures, due to the clayey nature of the deposits.

The most successful methods attempted so far on a large scale have been hydraulicking, Snodgrass machines, and trommels. The old reports show that dredging methods attempted on the Catawba, Uharie rivers and Fishing Creek proved unsuccessful. Various reasons have been given for the failures of these dredges. The older inhabitants of the above sections state that the companies were unable to secure sufficient properties, others state that the dredges were not able to handle the clayey materials.

In such sections of the State, especially at the Portis and Parker mines, the abundance of plastic clays has made the recovery of the gold an impossibility. Several methods have been tried out unsuccessfully to disintegrate the clays. The clay is so tenacious that if trommels are used, the gold is so pulverized that it floats out in the clay slimes. The Snodgrass machines and log washers have also proved unsuccessful because the quartz fragments tend to prevent the revolving of the blades within the drums. After the clay has been thoroughly disintegrated by these methods, the gold is worn so fine that it floats out in the slimes and does not come in contact with the amalgamation plates.

Two problems will have to be solved before many of the placer deposits can be operated profitably in North Carolina. The first problem is the disintegration of the clay without the pulverizing of the gold to such a fineness that it floats out in the slimes. The second problem is the recovery of the gold from the clay slimes at economical cost regardless of the fineness of the gold. If these two problems can be solved and the ore can be handled at low cost, there are many placer and saprolite deposits in the southeastern United States which could be worked profitably. A process of disintegration will have to be worked out in which the gold will be eliminated in the very beginning so that it will not be pulverized to such a fineness that it will be impossible to recover it. After the clay has been completely disintegrated, some process of recovery will have to be devised to recover the fine gold from the clay slimes. All processes so far attempted have been failures due to the inability of the operators to recover the fine gold by straight amalgamation and on English blankets, as the clay, more or less in a colloidal state, coats the plates in such a manner that the gold never comes in contact with them. Also the burlap blankets and English blankets become so covered with the fine clay that the gold floats off in the water.

Index

www.ingramcontent.com/pod-product-compliance
Lightning Source LLC
Chambersburg PA
CBHW050121210326
41519CB00015BA/4048